Science, Technolo
and Social Progr

Se[ries]

ST[
ST[

Edi[

AL[
E[

PAU[
U[

MEL[
GE[

EDW[
UN[

DON[
UN[

CARL[
POL[

JEROM[
UNI[

JOAN[
UNI[

LANCE[
WOR[

Science, Technology, and Social Progress

Research in Technology Studies,
Volume 2

EDITED BY

Steven L. Goldman

Bethlehem: Lehigh University Press
London and Toronto: Associated University Presses

Associated University Presses
440 Forsgate Drive
Cranbury, NJ 08512

Associated University Presses
25 Sicilian Avenue
London WC1A 2QH, England

Associated University Presses
P.O. Box 488, Port Credit
Mississauga, Ontario
Canada L5G 4M2

The paper used in this publication meets the requirements of the American National Standard for Permanence of Paper for Printed Library Materials Z39.48-1984.

Library of Congress Cataloging-in-Publication Data

Science, technology, and social progress.

(Research in technology studies ; v. 2)
Contents: Technology and international competitiveness / Christopher T. Hill — Technological frontiers
and human integrity / Langdon Winner — Automation
madness, or the unautomatic history of automation /
David F. Noble — [etc.]
1. Science—Social aspects. 2. Technology—Social
aspects. I. Goldman, Steven L., 1941– . II. Series.
Q175.55.S293 1989   303.4'83   88-45718
ISBN 0-934223-05-X (alk. paper)

Contents

6   Contents

Foreword

STEPHEN H. CUTCLIFFE
STEVEN L. GOLDMAN

This volume is the second in the series *Research in Technology Studies* published by Lehigh University Press under our general editorship. The intention of the series is to present a representative range of scholarship bearing on issues engaging the community of researchers studying the relationships between science, technology, and society. Each volume in the series focuses on a topical theme and consists of essays invited by the guest editor(s) of the volume. Our aim is to make the fruit of this contemporary research available to a wide audience interested in the social relations of science and technology.

The relationship of science and technology to social progress is the focal theme for the essays of this volume. Although the linkage of progress to the rise of modern science and technology has been a longstanding popular commitment, criticism by intellectuals of the idea of progress suggests that this linkage deserves a fresh examination. We are pleased to present a spectrum of views by leading scholars on the notion of science and technology as internally progressive and as sources of social progress. These essays should contribute to a clearer conceptualization of the idea of progress as it relates to developments in science and technology.

Introduction

STEVEN L. GOLDMAN

The twelve essays in this volume were originally presented at Lehigh University's Regional Colloquium in Technology Studies during the academic year 1986–87, supported by the Andrew W. Mellon Foundation. Under the joint sponsorship of Lehigh's Science, Technology, and Society Program and its Technology Studies Resource Center, colloquia were devoted to "Technology and the Ideology of Progress" (27 October 1986), "Science and Social Progress" (7 February 1987), and "A Critique of the Idea of Progress" (1 May 1987). With the assistance of the GTE Foundation, each of the colloquium meetings was preceded by a public lecture on a related theme: Christopher T. Hill of the Congressional Research Service on innovation and competitiveness, Nobel Laureate Rosalyn S. Yalow on radiation policies, social historian Christopher Lasch on the idea of progress today.

The quality of the presentations, their thematic coherence, and the many instances of mutual illumination urged formal publication in a single volume. This marks a break with the practice of separate publication of each colloquium's proceedings in a soft-cover Working Papers format. At some sacrifice in timeliness and cost, the advantage was gained of offering to a broader audience a wider range of perspectives on an important theme in science, technology, and society studies. The essays are grouped by colloquium topic and follow one another in the chronological order of their original presentation.

So much for the form of the book. The content is a mosaic of critical reflections, not so much on the idea itself of progress, as on its remarkable persistence. If for humanities intellectuals the idea of progress remains the "deadest of dead ideas" that Lewis Mumford declared it to be half a century ago, or is "now a concept unfortunately so unfashionable as to seem almost ridiculous" as the historian Lawrence Stone recently described it, it is nevertheless an idea that is alive and apparently well within the practitioner communities of scientists, engineers, technologists, science and technology journalists, political and corporate science and technology policy makers, and the public at large, at least in its interactions with science and technology. In one way or another, all of the contributions to this volume are responses to the persistence of the notions that science and technology are internally progressive and are sources of social progress as well.

David F. Noble, Langdon Winner, Joan Rothschild, Sal Restivo, and John M. Staudenmaier, S.J., all take sharp exception to the claim that science and technology, as currently constituted, can be socially progressive. Christopher T. Hill and Daryl E. Chubin focus on two of the linkages between society and the internal progress of science and technology. Carl Mitcham refuses to concede that science and technology *are* internally progressive and Christopher Lasch insists that what science and technology have to offer is not central to social progress as traditionally conceived. Kristin Shrader-Frechette, Rosalyn Yalow, and William B. Provine accept that science and technology are internally progressive and argue, in very different but clearly related ways, that science and technology are socially progressive in principle, but that obstacles currently exist to the expression of this character.

Yalow and Provine take the narrowest perspectives on the theme of progress, which they treat implicitly in the course of discussing societal responses to particular scientific developments: radiation physics in the case of Yalow and evolutionary biology for Provine. For Provine, scientific theories reveal the way things are, in spite of historical and sociological factors that enter into their formulation. Darwinian evolutionary theory has revealed to us that nature is in an important sense mechanistic. This implicates our "sense of meaning in life, our conception of our place in nature, indeed the whole foundation of ethics, including the question of whether we can make choices freely." It undermines deep-rooted Western cultural notions of purposiveness and progress that are fed by both the Greek rationalist-intellectual and the Judaeo-Christian religious traditions. As conceived by Darwin and his followers, evolutionary change is a "totally purposeless process, one that is completely opportunistic, with no sense in it of a larger design of any sort."

None of this, Provine insists, implies that life is meaningless or that one human behavioral choice is no better than any other. It does imply that human meaning and value can only be grounded in human experience; but rational norms of behavior can be derived from personal experience and the citizens of a post-Darwinian society should be engaged in the task of explicating these norms. This would constitute a progressive social development, one cognate with the expansion of our knowledge of biological nature. That this is not the case today is a consequence of the effectiveness of threatened social institutions and values in preventing the assimilation of the lessons of evolutionary theory and its human implications. Biological science thus offers the knowledge humanity needs to take the next step on its way to maturity, but established forces in our society will not let that step be taken. Yalow argues that an analogous situation exists with respect to the benefits offered by nuclear-physics-based technologies.

Yalow's Nobel Prize in physics was awarded for her pioneering research in radioimmunoassay techniques and their many applications, especially medical ones. Radiation physics over the past forty years has given rise to a number of technologies that are of unequivocal benefit to society. Yalow's concern is that this benefit is today threatened "by what has become an almost phobic fear of exposure to radiation at any level". She reviews the data bearing on such a fear, moving from a brief discussion of radiation exposure standards to an examination of the evidence—from atmospheric nuclear weapons testing (and use), diagnostic and therapeutic exposure to radiation, and radiation in the workplace and home—that low levels of radiation pose a significant threat to society. Her conclusion is that they do not.

All of this is intensely controversial, of course, particularly the presentation and interpretation of statistical data. The biological significance of short-term and of long-term exposure to low levels of radiation has been in dispute even among specialists for decades. The value of Yalow's essay lies not in resolving this dispute within science, but in calling our attention to a corollary that bears on the social relations of science and technology generally. Yalow writes, with specific reference to decisions to use, regulate, or forgo using available radiation-releasing technologies, that "public policy decisions need to be made in the absence of scientific evidence. It should be appreciated [by the public] that these are arbitrary decisions based on philosophy, not fact, and may well change because of political or other [nonscientific] considerations."

This is not a situation unique to radiation-related technologies. All science and technology policy decisions, indeed all policy and all personal decisions, must be made in the absence of "scientific evidence" that can objectively determine the one, correct, choice to make. What Yalow's analysis underscores for us is that the very notions of fact and of evidence can themselves be part of what is in dispute in scientific controversies. Every decision to act, then, even on science- and technology-related issues, is ultimately arbitrary in the sense that it is determined by adding to a body of inconclusive evidence a number of value judgments that are inevitably contingent and contextual. This leaves the decision open to challenge. It also entails that the social impact of science and technology is a consequence of value judgments.

The claim that the social impact of science and technology is progressive is a further value judgment, reflecting in part the response of those groups within a society that have the power to determine that science and technology are to be cultivated in one way rather than another. It also reflects the success of these groups in convincing other groups of the validity of their judgment. This is the theme of the essays by Noble, Winner, Rothschild, Restivo, and Staudenmaier. Each of these writers identifies a particular set of values, a different set for each writer, that shapes the conduct, directions, and evaluation of the outcomes of scientific research and technological innovation. For Noble, technological innovation in the West is driven by a managerial "madness" for control of work, the workplace, and workers. This madness manifests itself in the form in which automated production is pursued: with the objective of achieving production, and thus "progress," without people. Modeled after the military command hierarchy, this "lust" for control flies in the face of a rational assessment of the production process and the marketplace, let alone the well-being of workers or the public.

Because the citizens of Western societies believe in an "automatic future . . . driven and directed by an autonomous technological advance," technology can be cultivated in the way that management chooses and still be proclaimed progressive by the public. The public believes that a kind of Darwinian selection pressure acts on new technologies. The free market, in an entrepeneurial, capitalistic economy, will winnow out innovations that are "unfit" on performance or economic grounds. But this belief is mistaken. Technological "advance" is neither automatic nor Darwinian. Technical critieria of excellence are regularly overridden by the managerial objective of control, the marketplace is always manipulated by vested interests, and the "hard-headed businessman" in fact often makes "soft" policy decisions, reflecting parochial class and institutional value judgments.

Noble concludes that it is far from clear that higher levels of automation in the production process actually lower the cost of production, as opposed to changing what is produced. Nor do we have as yet a clear understanding of the relation of productivity to competitiveness—or for that matter of what would count as an absolute measure of productivity—or of how the benefits of an improved competitive position are distributed within society. Nevertheless, in the name of competitiveness, management is today dramatically intensifying and broadening automation of the workplace, and even those segments of the public that bear the burden of dislocation largely support this as a necessity if America is to "keep up."

Langdon Winner's essay echoes Noble's judgment that the dynamics of technological innovation are irrational, but locates the irrationality in social psychology rather than in class consciousness. The American public's apparently unbounded support for technological innovation as inevitably progressive reflects a "not entirely rational" fascination with technological advance as an end in itself, a fascination "rooted in colorful dreams, myths, and fantasies" that overwhelms every effort to assess technology critically.

Winner traces this fascination to the persistent identification in American culture—from the earliest Pilgrim settlers to advocates of space colonies and the Strategic Defense Initiative—of "the frontier" as the locus of human prospects for well-being. For Puritan Divines, the frontier was spiritual, the antechamber of salvation. Successively, the frontier was redefined as a secular arena for material prosperity (a transition described in Staudenmaier's essay), as a physical boundary (in the manner of Frederick Jackson Turner's study), as the unlimited domain of expanding scientific knowledge (Vannevar Bush's "Endless Frontier" and "Endless Horizon"), and today, amorphously, as "the future." But the future is just the name for not yet realized possibilities. Making the future the locus of human well-being deflects attention, concern, and a sense of responsibility away from the present, which is, in any event, superceded as we talk about it.

The result is that even technologies that seem to threaten our essential integrity as human beings—artificial intelligence applications, radical new computer-mediated human-machine interfaces, genetic engineering applications—escape critical control. They are pursued enthusiastically, with abandonment, as if we were Bacchantes pursuing that which, once seized, would assimilate the godhead to us. In the orgiastic pursuit, voices of caution are overwhelmed. Critics are stigmatized as naysayers actively impeding imminent salvation. This may press Winner's thesis a little hard, but Rothschild's essay, which focuses on the array of values underlying the commercialization of reproductive technologies, suggests that it is not being pressed too hard.

With the authors of all of these essays except Yalow and Provine (and possibly Shrader-Frechette), Rothschild takes progress to be an ideology, not just an idea. As such, it is freighted with a parochial sociopolitical agenda. The ideology of human progress, she argues, is rooted in beliefs in species hierarchy, especially the belief that humans are superior to all other species; in intraspecies hierarchy, that is, the belief that some humans are better than others; and in human perfectibility. New reproductive technologies are being developed and implemented not because of an "imperative" innate in the technical realm, and not for Winner's or Noble's reasons, but because they give us the power to act out antecedently held relevant values and the beliefs in which those values are given concrete form. These include "racist, classist, ethnocentric and patriarchal values" that motivate our interest in the new technologies and—calling Winner's central concern to mind—underlie our rush to exploit them in spite of what would appear to be manifest threats to traditional notions of human integrity.

Christopher Hill's essay is not concerned with ideology explicity, but it shares with Rothschild's the view that technological change is guided by antecedent values whose realization is promised by the new technologies. Technological change affects our lives by changing what we are able to do and at what cost. What we *want* to do, however, and the price we are willing to pay to do it, are value judgments that we already possess, and stand ready to renegotiate as technological innovations provide new opportunities to behave in relevant ways and at new costs.

Hill points to the incremental character of technological change as responsible for the acquiescence of a society in technology-abetted value changes: at each step along the way the new value (that is, the newly acted on, but "old" value) only seems to require a space in which to coexist alongside prevailing values. Very often, the new value then proceeds to devalue these, sometimes temporarily, sometimes permanently. In this way, even nonmarket values (privacy, body parts, motherhood, life) can be edged into the marketplace where, if commercial or political interests crystallize around them, they may be buffered from attack and given the opportunity to spread.

Recently, a new metaphor for progress has emerged in the West and promises to transform the value context of the social impact of technology. The new metaphor is "competitiveness" and it replaces previous metaphors that related social progress to the dominance of particular technologies. This transformation, which Hill believes is global, enduring, and important, implies a new politics of technology coordinate with the values that underlie competitiveness. Unlike the old politics that was based on *exploiting* comparative advantages, conferred by the possession of particular technologies, the new politics will be based on *creating* comparative advantages. Hill locates this shift historically, noting that it is a response to the fact that "key advances in bellweather technologies" are occurring elsewhere than in the United States. After the period from 1965 to 1980 when American society's consciousness of technology seemed dominated by technology's "negative externalities," and after the technological expansionism of the first Reagan term, we have begun to reconstruct the value structure of the society-technology relationship. We can expect that the consequences for the way we live will be profound, though we cannot yet specify what they will be. Once they are actual, however, it will be apparent (retrospectively) that this reconstruction proceeded on the basis of values whose realization was made commercially, politically, or personally attractive by the new relationship. Technological innovations will have created neither the new values, nor the new relationship. The innovations will have motivated the crystallization of new combinations of existing values, the actualization of which entails modifying the social, political, and economic institutions aligned with the prevailing value combinations.

The upshot of Hill's analysis, then, is that the social impact of applied science and technology is determined by the constellations of values that take shape in a society at any given time. This would appear to be a fundamentally political process, involving the relationships of power operative in that society, and it links Hill's essay to those of Noble, Winner, Rothschild, Restivo, and Staudenmaier in spite of the absence in it of explicit reference to ideology.

Rothschild's treatment of the ideological structure of the relation of science and technology to society is restricted to the ideology underlying the notion of human progress. Noble and Winner focus on class and social consciousness as determinants of the directions in which technology is led to push society. In their essays, Restivo and Staudenmaier offer broad cultural critiques of the social relations of science.

Restivo distinguishes science-in-principle, that is, science as it might be, and science-as-it-is, namely, modern science. Modern science is a social institution of a particular sort, linked to technology, to ruling elites, to state interests, and to the God who guarantees the prevailing moral order. (Provine amplifies this last point in his discussion of the anxiety of the scientific community today to avoid the equation of science and atheism.) Modern science is "intensely coupled to structures of class and power." In combination with the institutionalized ideologies of rationality and progress, modern science "is at the root of our social and environmental ills rather than an unambiguous force for freedom, understanding, and social progress."

Science-in-principle, however, *can* be an unambiguous force for freedom, understanding, and social progress. The notion of "pure" science is purely mythic, because "self and mind are social strategies," but Restivo claims that an epistemic strategy exists that will escape the "twin demons" of relativism and irrationality, and generate a nonideological, socially progressive science. Social and epistemic progress can be collateral consequences of wedding "anarchy" to "human inquiry." By anarchy, Restivo means a radical extension of participatory democracy, joined to the rigorous implementation of an antiauthority principle. The term "human inquiry" is less well defined, but it incorporates a methodology, based on the scale and scope of past and future scientific and technological "successes," that seems pragmatist, but which Restivo claims nevertheless avoids the relativism commonly considered a characteristic of pragmatism.

The result of human inquiry, as Restivo describes it, is "vertical social progress," the spread of freedom from want and authority, and of the opportunity to develop and to exercise creative and critical talents. Modern science is at best able to generate only "horizontal" social progress. That is, the spread of modern science's social benefits is restricted by the class structure of its social institutionalization, which affects, and infects, its methodology and its epistemology no less than its relations to prevailing extrascientific values.

Like Restivo, Staudenmaier also acknowledges that science has an innately progressive social character that is inhibited by an embedded ideology, but the ideology Staudenmaier perceives as operative within modern science is different from the one Restivo perceives. For Staudenmaier, progress talk in Western culture is ideological through and through. In its modern form, progress talk was born out of the convergence of science, technology, and economics in the late eighteenth century. Its ground, in spite of episodic reinterpretations of the state toward which progress is carrying us, remains the Cartesian disjunction of context and method.

The seventeenth-century scientific "revolution" incorporated the Cartesian disjunction in the form of a method for generating a body of knowledge that was value free and thus purely objective. Modern technology, increasingly science-based, incorporated the disjunction in the form of its characterization as a set of purely technical capabilities, intrinsically value-neutral. Capitalism, primarily after Adam Smith, was the economic image of value-free scientific theory and value-free technological practice. In time, it generated its own value-free managerial science and econometric models for value-free economic policy decisions. Taken together, science, technology, and economics contain the values that constitute the ideology of progress, which "has operated as a quasi-religious belief" in Western societies in recent centuries, but "today needs to be critically examined."

In fact, progress talk fails on logical, historical, and moral grounds. The claim that knowledge and technique are value-free cannot stand critical examination. The recent history of Western societies hardly supports the claims of the prophets of progress. The behavior of Western societies toward one another and toward non-Western societies in the modern era cannot escape moral condemnation. And yet, progress talk, and belief in progress, particularly in science- and technology-based progress, persists. This is all the more striking in light of the requirement—noted by Mumford more than half a century ago—of human passivity demanded by contemporary technologies. If societies want to benefit from advanced technologies, then people must submit to the instructions flowing from basic and applied science, as communicated by technical experts. The rule of reason encourages conformity, not individuality; submission to expertise, not independence; subordination to the system, not freedom to be idiosyncratic.

Staudenmaier uses the Chicago International Exposition of 1937 to support his claim that passivity was perceived to be the price of technology-based progress. The motto of the Exposition, presented to visitors as the markings over the gates of Hell were presented to Dante and Virgil, was "Science Finds; Industry Applies; Man Conforms." The intended lesson was echoed in a statuary group at the entrance to the main hall, consisting of male and female figures being shooed along by an outsize robot whose "hands" were at their backs. An accompanying text explained that the robot was urging the man and the woman "on to a fuller life." This relates, surely, to Restivo's emphasis on "anarchy" as one of the conditions of "true" science and on freedom from authority as one of the characteristics of the "vertical" social progress that will flow from the cultivation of true science.

Where Noble, Winner, Rothschild, Restivo, and Staudenmaier attack modern science and technology, together with progress, objectivity, and rationality as ideological, Kristin Shrader-Frechette insists on the reality of scientific rationality and of "objective scientific progress." Without the latter as a realizable ideal, our daily decisions could only be based on "personal authority, partisanship, and coercion." The *rationality* of science is "guaranteed," for Shrader-Frechette, by the individual scientist's subjecting their theory construction to the criteria of explanatory power and predictive success. The *objectivity* of science is guaranteed by the functioning of the scientific community.

The key to solving the problem of providing a rational account of scientific progress is the nonarbitrary adjudication of conflicts over cognitive values. Shrader-Frechette first reviews a number of solutions proposed for this problem. She then accepts and extends Larry Laudan's three-level justification scheme in which disputes at the factual and theoretical level are to be resolved by referring to the level of methodological rules and values, and disputes at the methodological level by referring to cognitive values. Shrader-Frechette adds a fourth level: cognitive value disputes are to be resolved by reference to ethical and utilitarian values. Furthermore, this is not a simply hierarchical scheme; its levels are mutually interrelated in complex ways.

Cognitive values can be assessed rationally, that is to say, objectively, nonarbitrarily, using this modified scheme, by measuring them against such prudential considerations as risk conservatism and such utilitarian considerations as openness to innovative theories. (This recalls Paul Goodman's plea that engineering be understood as a branch of moral philosophy and Paul Feyerabend's argument for an anarchistic, ideally a Dadaist, inventiveness as the "true" scientific method.) Shrader-Frechette provides a concrete illustration of her thesis by analyzing in detail the Atomic Energy Commission decision to permit burying radioactive waste materials at Maxey Flats, Kentucky. The trenches were dug in shale formations that were supposed to be impermeable, yet Plutonium contaminants were found miles away, off-site, within ten years. This was thousands of times faster than predicted by the technical experts whose arguments had won the day and who continued to maintan that the contamination was not by migration through the shale. Their geology ensured that this *could not be* the case.

In her close analysis of the contending experts in this dispute, Shrader-Frechette makes a valuable distinction between two epistemic strategies: internal and external consistency. The former is the traditional notion of measuring scientific reasoning in a given branch of science against the prevailing state of that branch. This encourages insular reasoning, closing science in on itself. External consistency measures scientific reasoning both against internal scientific criteria and against the compatibility of the scientific conclusions reached with relevant extrascientific values. This strategy opens a wedge for bringing ethical and utilitarian considerations to bear upon cognitive values in a rational, rather than in an arbitrary, way.

Shrader-Frechette's logic proceeds from a commitment to the objectivity of scientific progress. Mitcham, by contrast, concludes, also with reference to Laudan, that the progressive character of scientific theory change is a value judgment at least as problematic as the claim that science is rational. Many philosophers of science, however, including Laudan, have occupied themselves with justifying this latter claim while taking the value judgment pretty much for granted. Mitcham opens his essay by noting that the "theory of progress is, despite the literature, fraught with philosophical issues still to be explored." He interprets the commonsense notion of progress as a "prereflective" identification of our own with what is good. The projection of this onto history leads to familiar linear story lines of the progressive maturation of science from Pythagoras to us. If, however, we adopt a critical posture and attempt to analyze historical change on the one hand, and scientific and technological change on the other, we find that, for all three, successive states can be judged superior to precedent states only ambiguously.

The problem boils down to being able to determine "betterness" objectively. Our patent inability to do this implies that progress "is not so much found in the history of science or technology as projected upon it." As with Kant, progress is an idea brought *to* experience, not one derived *from* experience. Progress is a historical and contingent idea. This is a deeper criticism of progress than exposing it as ideological. For Mitcham, the judgment that a particular state of affairs is progressive vis-à-vis some other state is ineluctably relativist, subjective, contextual. No reconstruction of science—for example, one along lines that would satisfy Shrader-Frechette, Winner, Noble, or even Restivo—would affect Mitcham's defenestration of progress. And yet, ideology-riddled or critically vacuous though it may be, progress talk continues to play a central role in scientific discourse and in scientific practice, as the first part of Chubin's essay explains.

Progress is essential to the production cycle of scientific knowledge (and although Chubin does not discuss it here, to the production cycle of technological knowledge as well). Internally, "progress" functions as a means of relating the work of individual researchers to the collective enterprise in which they are engaged. It is a measure of the assimilation of group norms into the individual research setting. This dovetails nicely with Shrader-Frechette's claim that scientific rationality derives from the methodology employed by the individual scientist, while scientific objectivity derives from the social structures of the community of scientists. Progress internal to science is thus cognate with objectivity as an expression of the social structure of scientific practice. That is, internal progress is a judgment grounded in communal norms and values.

But progress talk also has another role in scientific discourse. It functions externally as well as internally, linking the conduct of science to its societal context. Internally, progress relates to the process of generating and estimating the value of scientific knowledge in relation to other scientific knowledge. Externally, progress relates to the products of the practice of science and to estimating their value to society. This poses something of a paradox. On the one hand, the scientific community has gone to considerable lengths to insist on the independence of "pure" science, of basic research. This claim has the effect of uncoupling science from society at the most fundamental level, so that there should be no expectation that scientific progress will be linked systematically to any other aspect of social life than intellectual sophistication.

On the other hand, in asking for societal support to conduct basic research, the scientific community promises practical contributions to social progress as a consequence of scientific progress. This has become especially true in the post–World War II period and above all in America, where billions of public dollars are annually given in direct support of basic research, and billions more indirectly. In return for their billions, the public is assured that it will benefit many times over in jobs, prosperity, security, health, and national prestige. For its part, the public, especially in the last twenty years, has asserted its right to a measure of accountability from science, regardless of science's past contributions to social well-being. Science responds by reaffirming its autonomy; in spite of which, it claims, public support must continue, if only for the public's own good. Ironically, science's very successes in serving social interests erode its autonomy by raising society's expectations.

The second part of Chubin's essay is devoted to an examination of the institutional mechanisms that shape the public's perception of science, in particular the media and science journalism. Science and the media are "two powerful professions, each seeking to exert control over and extract data from a source." On the face of it, the source for the scientist is nature and for the journalist, some person or group within the scientific community. But Chubin penetrates behind appearances to the reality of the situation. The tension between public accountability and professional autonomy that is the price of public support makes it the case that the scientist whose research is newsworthy attempts to control the medium through which the public will learn of that research, and of its actual or potential value. The scientist and the science journalist then, often (obviously not always) are engaged in wrestling for control over one another, because each has professional objectives embodying distinctive relations to society.

Provine touches on this when he writes of the anxiety of the scientific community that scientific knowledge not be perceived as entailing atheism, for fear of alienating Congressional support for science. Just so, scientists "play up" to science journalists in order to get the publicity that will win the grant support they need. The upshot is that the public's perception of science is skewed and science journalists who attempt to do a critically respectable job are liable to be perceived as hostile to science, self-serving careerists, or muckrakers. Chubin concludes that the stereotypes of science and scientists do a disservice to a public increasingly activist with respect to influencing the setting of science and technology policies. What is needed is for science reporting to focus on the process of the production of scientific knowledge, especially on its internal sociology and on its relationship to the societal context within which science is practiced.

Chubin thus softens his critique of the science-progress-society relationship by suggesting that the flow of socially progressive benefits from science is only circumstantially impeded. With a deeper understanding of the process of science—an understanding to be communicated by a more sophisticated science journalism—society could construct a more enlightened relationship to science. Similarly, Yalow and Provine place the onus on society for not taking up pearls that science has cast before it, while Noble, Winner, Rothschild, Restivo, and Staudenmaier argue that scientific research and technological innovation happened to be practiced in invidious ways reflecting the operation of corrupting sociopolitical ideologies.

In principle, though with varying degrees of difficulty, these ideologies could be changed and, again in principle, appropriate changes would generate new, socially progressive forms of science and technology. Shrader-Frechette, too, offers a solution to the problem of optimizing the innately progressive character of science by opening up the scientific method in the way she works out: linking cognitive value disputes to ethical and utilitarian values by way of the epistemic criterion of external consistency. Hill alone of the authors here avoids judging progress. He is satisfied to describe the recent transformation of the technology-progress-society relationship and the socioeconomic mechanism by means of which new technologies affect our lives. Mitcham stands out for explicitly refusing to concede the reality of internal scientific progress, but Christopher Lasch goes a step further in proclaiming the irrelevance of science and technology to progress.

The eighteenth-century belief in progress was a belief in a cumulative and irreversible historical pattern of global improvement, in the existential circumstances of human beings to be sure, but far more significantly, in human nature itself. It needs no arguing at this moment in human history that the latter is not yet discernible, in spite of extraordinary increases in technical knowledge and technical mastery over the circumstances of human existence. Science and technology, therefore, simply are not indices of essentially human well-being. To Lasch, "fear and trembling are a more appropriate response to the [scientific and technological] facts at our disposal than a heightened confidence in our powers."

Freud erred when he argued that believing in the truths of science rather than in the teachings of religion would be a personally and socially progressive development, a maturation for humanity. Lasch interprets the beliefs in mastery and freedom associated with science and technology as deeply rooted in a denial of human dependency on forces beyond human control. Embracing science and technology as sources of power and unlimited possibility is a regression to postinfancy feelings of omnipotence. Outgrowing such feelings really *would* be a sign of personal maturation. Furthermore, underlying the popular Western belief in science-based progress is a belief that the emotions stand in opposition to the intellect. The emotions, driven by desire, which is in turn driven by physical appetites, occlude clear thinking, which transcends bodiliness. In the context of such a belief—in the promotion of which Descartes is again inculpated, as he was in Staudenmaier's essay for the harmful disjunction of method and context—it is possible to anticipate with approval that humanity will one day outgrow the need for a body, thereby finessing the senses, desire, and irrational passions.

Even on the face of it, however, how likely is it that an interpretation of humanity that is hostile to the life of the senses, to the body, to emotion, and to desire is capable of undergirding a set of activities—scientific research and technological innovation—whose consequences will improve the human condition? With this in mind, Lasch finds the arrogance of the modern scientific-technological age "depressing." What is needed if human beings are to become better as human beings is that "a spirit of humility and contrition" be brought to the interpretation of our experience, not the swagger of an adolescent fantasy of playing Titan.

Fifty-six years ago, J. B. Bury began his classic study by identifying the "idea of the progress of humanity" as one of those ideas "that play a great part in determining and directing the course of man's conduct but do not depend on his will—ideas which bear upon the mystery of life, such as Fate, Providence, or personal immortality. Such ideas may operate in important ways in the forms of social action, but they involve a question of fact and they are accepted or rejected not because they are believed to be useful or injurious, but because they are believed to be true or false." Bury's critical acuity is underscored by the twelve essays in this volume.

The continuing efficacy of the idea of human progress in shaping personal and social action reflects the persistence of a belief in the fact of progress, actual or potential. But this belief cannot be validated, Bury argued, because the fact of progress cannot be established. We do not know, we cannot know, whether all of the changes in society, or even in our science and our technology, are moving humanity "forward." "Certain features of our 'progress' may be urged as presumptions in its favor, but there are always offsets, and it has always been easy to make out a case that . . . the tendencies of our progressive civilisation are far from desirable." Thus, if we agree that personal happiness for all would constitute unambiguous human progress, we have no way of knowing that civilization is in fact moving in that direction, though some would interpret appearances in that way. Nor is there warrant for the assumption that science and technology will continuously advance either to a complete knowledge of nature, or even to "an immeasurably larger and deeper knowledge than we at present possess." And the belief in our continuing moral and social perfectibility "rests on much less impressive evidence."

In short, the idea of social, intellectual, and moral progress, like the beliefs in Providence and immortality, rests on an act of faith. But the belief in progress, in establishing itself in modern Western culture, had to overcome a profound psychological obstacle, one Bury calls "the illusion of finality." This illusion leads us to interpret our own situation in cultural space-time as coming at the head of a "progressive" historical development, in the manner

Mitcham discusses in his essay. Science has been our primary aid in overcoming this illusion and permitting us to imagine that our descendants will live in a world as fundamentally different from our own as we recognize our own as being from our ancestors'. But, Bury concludes, once we acknowledge a historical perspective on human development, can we exempt the idea of progress itself from this process? Must we not expect that in time "a new idea will usurp [progress's] place as the directing idea of humanity"? This new idea will then be "the criterion by which Progress and all other ideas will be judged. And it too will have its successor." Paradoxically, then, the historicity of the idea of progress, whose personal and social force lies in believing that it is a fact, hence absolute, would appear to lie in its being relative to a "certain not very advanced stage of civilisation."

Have Western societies progressed since the rise of modern science and technology? If it is claimed that they have, have science and technology played an essential role in causing that progress? Western societies have certainly changed in the past three hundred years and science and technology are unequivocally implicated in causing many of these changes, most obviously those associated with production, consumption, transportation, communication, energy, war, and health. But do these changes constitute absolute improvements in the human condition? The temptation is to answer "yes" and to support this by pointing to morbidity and mortality statistics; the spread of education; the demise of traditional slavery and serfdom; dramatic changes in the social status (at least in principle) of children, women, and minority groups, especially blacks; the expanded sphere of individual rights against state and vested interests; increases in real income; and changes in housing, diet, sanitation, and social services. And, of course, science and technology themselves certainly seem to have improved almost continuously, as bodies of knowledge and of techniques, by comparison with their earlier states.

Against this must be ranged the horrific violence and cruelty, the profligacy and mindlessness of "advanced" Western societies, most notably over the past one hundred years; the persistence of prejudice, oppression, corruption, and militarism, all of which have found means of exploiting science and technology to their own ends; and the argument that all of the changes pointed to as positive affect only the externalities of human existence while the negatives reveal that human nature is at best unchanged. The controversy, as noted in relation to Mitcham's essay, can only be resolved by identifying a nonarbitrary basis for resolving the asymmetry between "change" and "improve." Every claim of improvement entails a change having been made, but not every change is an improvement. If we ask, "How do we judge when a change is an improvement?" we discover a fundamental distinction between artifacts and "naturefacts," that is, that which we have not deliberately made.

With regard to artifacts, it is in principle possible to determine if a change is an improvement because every change can be measured against the purposes for which the artifact was made. Granted, these purposes, especially early in the history of an invention, can be poorly formulated and they may change as new applications occur. Yet, as a general rule, we can always specify the context relative to which a claim that an artifact has been improved is to be measured. No such context can be specified for naturefacts except arbitrarily. We cannot say what animals, rivers, the atmosphere, the moon, asteroids, or human beings are for. Therefore, changes wrought in these can only be claimed to be improvements relative to the intentionalities underlying human action. Breeding chickens so that they lay more eggs or grow plumper thighs for human consumption; using the oceans, rivers and atmosphere as sinks for human pollutants; mining the moon and asteroids for raw materials for production needs; changing humans so that they will be more productive members of society and help their societies to compete with other societies: none of these can be judged improvements in the entities affected without an antecedent determination of what those entities are for. And, of course, it is just this determination that is necessarily lacking.

For another reason than he gave, then, Bury's insight is vindicated. Claims of human progress must remain relative to value judgments, concerning the basis and end of human existence, that are necessarily nonunique. They will, as a result, always be challenged by competing visions of human beings. Such claims must also be sharply distinguished from claims that science and technology, at least, are "truly" progressive. Indeed, the term "progress" is used equivocally in referring to human beings, technology, and science. Technology unquestionably progresses relative to specifiable performance objectives, but the human consequences of new artifacts do not "inherit" their progressive character vis-à-vis other artifacts. For example, computer technologies can be said to progress as increasingly powerful and competent computers are produced at lower cost, where "powerful" and "competent" can be defined relative to unambiguous performance specifications such as memory size and speed of computations. The social consequences of increasingly powerful yet vastly cheaper computing power are, as we are coming to understand, profound. Yet these changes in human values, relationships, and behavior patterns cannot be called progressive just because the technology precipitating changes is progressive.

Turning to science, the claim that it progresses as its theories become more powerful can only be defended if it is supposed that scientific theories are truer the more closely they conform to what exists "out there," independent of human experience. But that, of course, is just what we cannot know. The predictive successes of science, the new technologies invented on the basis of scientific theories, cannot establish that any given scientific claim conforms to the way things are. Furthermore, the philosophical, historical, and, more recently, the sociological critiques of science as socially "constructed" constitute formidable rebuttals of classical realist claims that science objectively, that is, impersonally, reveals the structure of reality. From a critical perspective, then, scientific progress becomes conflated with technological progress, though there is little doubt that most scientists would find this nonsensical.

The essays offered in this volume remind us that the notion of progress, because it remains on so many lips, deserves to be treated thoughtfully and needs to be, because of its complexity.

Contributors

DARYL E. CHUBIN is Senior Analyst in the Science, Education, and Transportation Program of the Office of Technology Assessment of the U.S. Congress. He is the Consulting Editor of two new journals, *Knowledge in Society* and *Technology Analysis and Strategic Management,* and was Project Director for the OTA report *Educating Scientists and Engineers: Grade School to Grad School.*

STEVEN L. GOLDMAN is Andrew W. Mellon Distinguished Professor in the Humanities at Lehigh University and a member of the History and Philosophy departments. He is the former director of Lehigh's Science, Technology and Society Program. His research focus is the social relations of science and technology in Western culture.

CHRISTOPHER T. HILL is Senior Analyst in science and technology policy for the Congressional Research Service of the Library of Congress. He is the author of *Technological Innovation for a Dynamic Economy* (Elmsford, New York: Pergamon Press, 1979), and co-author of *Regulation, Market Prices and Process Innovation: the Case of the Ammonia Industry* (Boulder: Westview, 1979).

CHRISTOPHER LASCH is the Watson Professor of History at the University of Rochester and chairman of the Department of History. His publications include: *Haven in a Heartless World* (New York: Basic Books, 1977); *The Culture of Narcissism* (New York: Norton, 1979); and *The Minimal Self* (New York: Norton, 1984).

CARL MITCHAM is a Professor of Philosophy and Director of the Philosophy and Technology Studies Center at the Polytechnic University (New York). He is the co-editor of *Philosophy and Technology* (New York: Free Press, 1972; repr. 1983) and of *Theology and Technology* (Lanham, MD: University Presses of America, 1984).

DAVID F. NOBLE is a Professor in the Department of History and Politics at Drexel University. He is the author of *America By Design: Science, Technology and the Rise of Corporate Capitalism* (New York: Knopf, 1977) and

Forces of Production: A Social History of Industrial Automation (New York: Knopf, 1984).

WILLIAM B. PROVINE is Professor of the History of Biology in the section of Ecology and Systematics, Division of Biological Sciences, and in the Department of History at Cornell University. He is the author of *Sewall Wright and Evolutionary Biology* (Chicago: University of Chicago Press, 1986).

SAL RESTIVO is a Professor of Sociology and of Science/Technology Studies and Director of Graduate Programs in the Department of Science and Technology Studies at Rensselaer Polytechnic Institute. He is the author of *The Social Relations of Physics, Mysticism, and Mathematics* (Holland: D. Reidel, 1983), and co-editor, *Comparative Studies in Science and Society* (Columbus, OH: C.E. Merrill, 1974).

JOAN ROTHSCHILD is a Professor of Political Science at the University of Lowell (Massachusetts) and had been a visiting scholar at the Philosophy and Technology Studies Center at Polytechnic University (New York). She is the author of *Teaching Technology from a Feminist Perspective: A Practical Guide* (Elmsford, New York: Pergamon, 1986) and the editor of *Machina Ex Dea: Feminist Perspectives on Technology* (Elmsford, New York: Pergamon, 1983) and of *Women, Technology and Innovation* (Elmsford: Pergamon, 1982).

KRISTIN SHRADER-FRECHETTE is Graduate Research Professor of Philosophy at the University of South Florida. Her special interests include the philosophy of high-energy physics, philosophy of economics, science/technology policy, environmental ethics, and, most recently, philosophy of geology. She is the author of *Science Policy, Ethics, and Economic Methodology,* (Holland: D. Reidel, 1984), and *Risk Analysis and Scientific Method,* (Boston: Reidel, 1985).

JOHN M. STAUDENMAIER, S.J., is Associate Professor of the History of Technology at the University of Detroit. His *Technology's Storytellers: Reweaving the Human Fabric* (Cambridge, Mass.: MIT Press, 1985) received the 1986 Alpha Sigma Nu Award (National Honor Society for Jesuit Colleges and Universities) for best book in the social sciences.

LANGDON WINNER is Associate Professor of Political Science in the Department of Science and Technology Studies at Rensselaer Polytechnic Institute. He is the author of *Autonomous Technology* (Cambridge, Mass.: MIT Press, 1977) and *The Whale and the Reactor* (Chicago: University of Chicago Press, 1986).

Rosalyn S. Yalow is a Nobel laureate, a member of the National Academy of Sciences, the American Academy of Arts and Sciences, and the recipient of forty honorary doctorates. She is Senior Medical Investigator at the Bronx (New York) Veterans Administration Hospital, and Solomen A. Berson Distinguished Professor-At-Large at the Mt. Sinai School of Medicine (New York).

Science, Technology, and Social Progress

Part I
Technology and the Ideology of Progress

Technology and International Competitiveness: Metaphor for Progress

CHRISTOPHER T. HILL

The dominance of particular technologies is often used to mark the "progress" of modern Western societies, expressed by labels such as the Ages of Steam, the Automobile, Atomic Energy, Information, Space, and Bionics. For the first time since the Age of Steam, America faces the possibility—indeed, the reality—that a significant number of key advances in the bellweather technologies of our era are occurring somewhere else. "Progress," or the lack of it, is now being marked by the degree of our success in competing for international markets in "high-technology" goods and services. Have we entered the "Age of Competitiveness"? If Progress is marked by Competitiveness, what do we mean by it? How can we achieve it? How might seeking to become more competitive affect our lives and our society?

In discussing these issues, I consider three aspects of progress. First, I offer some general comments on how changing technology affects our lives and the values we hold, and I relate these changes to our notions about "progress." Second, I argue that during the last decade, we have experienced a fundamental shift in how we view the "problem of technology" in American political discourse. The focus of national attention has shifted from managing the undesirable consequences of rapid technological change and promoting certain technologies thought to be in the "national interest," to encouraging more rapid technological development and change across a broad front in order to meet the challenges of international competitiveness. Third, I describe some of the interests and activities now underway in the U.S. Congress that illustrate the nature and direction of this new interest in technology and international competitiveness as "metaphor for progress."

TECHNOLOGY AND PROGRESS

The development, application, and control of new and better technologies play profound roles in human progress. Here, "progress" means not only material progress, by which new and better products help to meet human wants and needs more effectively, and through which better processes enable us to satisfy those wants and needs using less of nature's resources, but also the more satisfactory achievement of the less tangible goals we value highly as individuals, as a society, and as a culture. These include security, independence, freedom, production, communication, health, pleasure, autonomy, control, longevity, community, justice, variety, creativity, and on, and on. Indeed, in every culture, no matter the material level and the set of values adhered to, it is in the nature of being human to use tools—that is, technology—in seeking to achieve these kinds of highly valued goals.

Paradoxically, even as new technologies contribute to progress, they can cause unwanted changes and material harm, which are experienced not as progress but as loss. Indeed, such technologies as nuclear and biological weapons can threaten life itself. Here, material harm means undesirable physical damage to human life, to the physical and biological environment, and to material goods. It includes such diverse outcomes as environmental degradation, the waste of natural resources, the destruction of peace and quiet by noise and overcrowding, and the desecration of sacred regions and relics.

Technological change can also cause profound changes in highly valued nonmaterial aspects of life. For example, some new information technologies threaten personal privacy, even as they make financial transactions cheaper and easier. New medical technologies bring the promise of more fulfilling life, while challenging our concept of what it means to live. Transportation technologies make it easier to move from place to place, but undermine our concept of community.

Thus far, I have discussed the implications for progress of some of the material and nonmaterial consequences of technological change while assuming that the underlying values remain constant. That is to say, I have implicitly judged whether a technological change is desirable or undesirable according to whether it supports or attacks the achievement of presumably fixed, fundamental values. In addition, however, technological change tends to change both the nature of underlying values and the intensity with which we adhere to them. That is, in addition to causing changes in the degree to which valued outcomes can be achieved, *new technology changes how we value the outcomes themselves.*

New technology changes values by changing both what we are *able to do,* and *how much it costs* to do the things we do. Though oversimplifying, we can say that new technology changes the possibilities and the costs of achieving, or of circumventing, valued outcomes, and thereby changes the values themselves.

How can this be? It seems paradoxical: we believe that values are not arrived at by weighing the costs and benefits of outcomes; they have to do with fundamentals—with things that are nonpriceable, cultural, or even spiritual, and certainly are not tradeable in markets. Only an extreme utilitarian would claim that everything has its price, yet over time and little by little, the radical changes in the costs and the possibilities of actions that result from technological change do affect the values we hold, and they can do so to a profound degree. In other words, the availability of lowcost means tends to change the ends to which the means are turned.

If new technology alters cherished values, why do we permit technology to change? One explanation can be found in the fact that not all individuals and groups adhere equally to the same set of values. Even if I object to the effects of a new technology, someone else may like what it does and decide to adopt it. However, unless my objection to the new technology ultimately prevails, my own preferences are likely to accommodate in some measure to the new circumstances the use of the new technology creates.

Perhaps an even more powerful explanation for the acceptance of technologies that create undesirable change can be found in the fact that, contrary to popular wisdom, technological revolutions that pose immediate ethical dilemmas are rare. Instead, technologies tend to evolve incrementally, and they create myriad everyday changes that rarely threaten. Value change arising from technological change may not be opposed early on when it is easy to do so, because the marginal effects of incremental technological change may not only not appear to be undesirable, but even appear to be benign at each step. When technological change is subtle and continuous, it is difficult to perceive it as a threat to values, and even more difficult to turn it aside. Each small change seems "better" on the margin, even though the aggregate change might seem profoundly undesirable if foreseeable at the outset. Furthermore, the continual value accommodation that occurs means that by the time the impacts of the change have been recognized, we have not only accepted it, we have embraced it.

Let me illustrate. How much is a child worth? Traditionally, not much. Before the advent of sanitation and medical care technologies, newborn life was precarious. Infant mortality was high, and the probability of surviving the diseases of childhood was low. It was not that it was expensive to raise a child to adulthood; it was often impossible. In such circumstances, deep parental and community attachment to any particular child was unwarranted. Instead, it was important to bear many children in the hopes that they would care for their parents as they aged, and that they would assure the survival of the community. Adults were people, children were not, and the latter were much less valued.

As disease processes became understood and low-cost disease prevention and medical care technologies were made available, infant mortality declined. Consequently, the disadvantages of emotional attachment to each newborn also declined, and each child's life could be more highly valued. Child abandonment became unacceptable in Europe in the eighteenth century, and child labor became unacceptable in the early twentieth century.

The evolution of medical technology continued, however, and techniques for ensuring infant survival steadily became more powerful. Methods of managing premature infants were developed to the point that now infants delivered many weeks early and weighing one pound or less survive. Fetuses with certain congenital abnormalities have been surgically treated in utero, and there have been recent reports of such treatments ex utero, with return of the fetus to the uterus after surgery. More and more "heroic" efforts are possible and executable at lower cost, and now newborn infants are thought of as having a "right" to life, a concept that barely existed a hundred years ago. Technology facilitated a reverence for each young life that millennia of religious reaching had not achieved.

Because each advance in life-saving technology has appeared to offer an advantage over the prior state of the art, it has received widespread acceptance. Each advance seemed to enhance the achievement of deeply held values, not to challenge the values themselves. Yet, in the aggregate, the advances not only established the right to infant survival (parental failure to provide for fetal and infant survival is now considered in law to be a serious breach of the infant's right to life), they also challenge us to cope with the new possibilities of fetal surgery, of fetal assessment and elective abortion, and of genetic manipulation (the latter feared in the abstract, but likely to be applauded in specific applications), as well as with the problem of assuring a decent quality of life for those infants that survive with severely diminished physical or mental capacities.

The cumulative effects of technological change on values are also illustrated by how new technology can lead to the reassignment of *nontradeable* goods to the category of *tradeable* goods, and vice versa. Tradeable goods are those that society is willing to price and to buy or sell for money, such as clothes, automobiles, machine tools, and theater tickets. Nontradeable goods are those valued things that most people do not feel it is legitimate to buy or sell in the marketplace, regardless of their prices, including such goods as liberty, privacy, votes, and human body parts. Sometimes the nontradeability of goods is enshrined in laws that prohibit market transactions involving them; in other cases, social mores provide the only control. For example, in modern American society, you cannot sell yourself into slavery at any price—your liberty is nontradeable.[1] Similarly, it is not legal to buy someone's vote, nor is it legitimate to sell your kidney or your left arm. On the other hand, while it is legal to purchase human blood from indigent donors, perhaps because blood can be regenerated, trading in blood is widely regarded as being on the fringe of legitimacy.

Current advances in medical technology are tending to shift some human body parts into the tradeable goods category. Although it has not been widely accepted as legitimate, there have been serious proposals in recent years to establish commercial repositories of body parts taken from the deceased to be sold for transplantation, and there have been reports of the purchase of kidneys from living persons for transplant. Such actions were inconceivable prior to recent advances in medical technology.

The new approach to surrogate motherhood based on artificial insemination or a contract mother with the sperm of the father has begun to shift parenting, or at least childbearing, from a nontradeable service to one that is sold in the marketplace: the typical fee today is ten thousand dollars, plus expenses. The challenge this practice creates for human values is indicated by the intense interest in the celebrated "Baby M" court case in New Jersey in which a surrogate mother has attempted to assert custody over a child she bore under contract. This example also shows how incremental change in technology can lead to major ethical and social dilemmas. After all, paying for the services of a surrogate mother is only marginally different from a more traditional practice in which prospective adoptive parents pay for the care of a pregnant woman who agrees to give her child up to them for adoption. And human artificial insemination is a straightforward application of a technology widely employed in animal husbandry.

The benefits of electronic information systems containing personal data are likely to become so large that our principled commitment to privacy may be substantially eroded by this new technology; people may essentially give up the "right to be let alone" in return for the time and money saved, for example, through electronic banking. This is somewhat ironic, as the medieval invention and adoption of a new technology, the chimney, made individual living and sleeping quarters possible, and this has been cited as one origin of the idea of personal privacy.

An example of how a good appears differently when it is treated as tradeable or nontradeable arises in the area of environmental quality. For at least two decades, this country has debated whether environmental pollution should be managed in a market or nonmarket framework. The debate turns on whether environmental quality should be viewed as a tradeable or nontradeable good. Those who would trade environmental quality usually prefer financial approaches to environmental management: pollution taxes or emission fees to discourage pollution, and investment tax credits or tax-free bonds to finance pollution-control investments. They hope to fine-tune these incentives and disincentives to achieve a "socially optimal" level of pollution, which is to be determined by weighing the costs and benefits of a clean environment against those of other valued outcomes like industrial production and jobs. In the extreme, they believe that polluters should pay for cleanup and control, but not that polluting is wrong.

On the other hand, those who see environmental damage as an assault on a nontradeable good—that is, on the *right* to a clean and healthful human and natural environment—characterize polluting as a *wrongful* act. They tend to prefer legal and regulatory approaches to controlling polluters. Rather than viewing polluters as people who are taking advantage of low-cost waste disposal opportunities, they view them as trespassers on the right to a clean environment. They prefer to use civil law suits and regulations with associated civil and criminal penalties to compel polluters to stop polluting, regardless of cost.

As Congress has adopted environmental laws over the last twenty years, it has shown some tendency to move closer to the market concept of pollution control. From the early mid-1960s to the early 1970s, pollution laws usually established regulatory standards based on public health considerations or on judgments about what could be achieved technologically, rather than on a balancing of the costs and benefits of control. More recent statutes, however, put greater emphasis on a balancing of environmental and other values, and interest has grown in using financial incentives and disincentives to control pollution. Combinations of regulatory standards and financial incentives have also been used, as in the coupling of regulatory performance standards for air pollution emissions with emission taxes for those who fail to meet the standards.

THE BIG SHIFT IN ATTITUDES TOWARD TECHNOLOGY

From about 1965 to 1980, America had an intense, ambivalent relationship with technology. Many Americans doubted whether the benefits of new technology were worth the costs. Business and government were widely seen as powerful forces that were imposing the undesirable human and environmental impacts of new technology on a nearly powerless citizenry. Although few were true "antitechnologists," many wanted to bring technology under closer scrutiny and stronger social control. Environmentalism, consumerism, "small is beautiful," and "technological unemployment" were bywords of the day.

At the same time, outstanding technical achievements like the Apollo moon landings led many to believe that the power of technology to solve important national problems was nearly without limit. In this belief, we turned to scientists and engineers to help solve a host of societal problems, from crime prevention and mass transportation, to energy conservation and weather modification.[2]

Thus, the national agenda for technology in the late 1960s and 1970s was highly visible and bifurcated: it featured initiatives to control technology and initiatives to harness technology to meet national needs. The high-water mark of this bifurcated attitude toward technology was reached in the mid-1970s with the passage of laws relating to energy supply and demand, culminating in the establishment of the Department of Energy in 1977.

The past ten years, and especially the past five, have witnessed a dramatic change in public and official attitudes toward new technology. This change has two parts. First, while the nation remains committed to controlling and managing the undesirable consequences of technological change, this commitment is no longer viewed as radical or antitechnological. Instead, technology assessment and the management of the side effects of technological change are routine elements of doing business in nearly every sector. Although arguments continue to be heard at the margin, for example, gross environmental damage is no more acceptable today than child labor or witchcraft.

The second change is that our attention has shifted from public programs that promote the development of specific new technologies intended to solve selected national problems, to public policies to stimulate development and use of diverse new technologies intended to address an overriding national problem—the decline in the technological competitiveness of American industry.

As a reflection of these changes, many members of Congress are more interested today in promoting a wide array of "high-technology" industries, which are expected to contribute to growth in productivity, competitiveness, jobs, and economic output, than they are in mobilizing national technical resources to address, say, garbage recycling, solar energy, or urban transit. This is not to say that the latter interests have disappeared, but only that the focus of attention has shifted away from these kinds of targeted programs and toward more systemic, wide-ranging policies that are intended to stimulate many industrial sectors.

These changes have important implications for the strategies, tactics, and management of public policies and programs related to new technology. They have also had significant influence on the relationship of government to business in America, on the relationship of universities to industry, and on the roles that such other interests as labor, consumers, and environmentalists play in public debates about technology.

Why have these changes in attitudes toward new technology occurred? Several factors are at work. Our declining balance of international trade in manufactured goods, especially in goods based on sophisticated technology, has been pointed to as one sign that our technology is not advancing as rapidly as it should. While arguments can be made that the overall trade balance is a consequence of macroeconomic forces beyond the control of technology (such as the overvalued dollar and high real interest rates), there is also evidence that the quality of American goods and the productivity of American industry have slipped compared with the performance of foreign competitors. Furthermore, that the decline in technological competitiveness predated the large changes in macroeconomic conditions that occurred between 1979 and 1982 has been shown by detailed comparisons of the quality and production costs of goods made here with those made abroad.

Another reason for the change in attitudes toward new technology is that many state and local officials have fastened on technology-based, entrepreneurial industry as the key to local and regional economic revitalization and development. The premier example is the "Route 128" area around Boston, where technology is widely believed to be the engine that drove the economic revitalization of this formerly depressed region. Now nearly every state has a highly visible program to attract or nurture technology-based industrial development.

A third reason for the change in attitudes is that technology has become an important vehicle for individual realization of the newly popular goals of material and financial success. The stories of Steven Wozniak and Stephen Jobs at Apple Computer, Bill Gates at Microsoft, and H. Ross Perot at EDS provide models of how technology-based industry offers enormous personal rewards.

And finally, the environmental and ecological disasters predicted by widely read authors in the late 1960s and early 1970s have not occurred. Contrary to expectations set up by such books as *The Limits to Growth*,[3] *The Closing Circle*,[4] and *Famine 1975!*,[5] the 1980s have not found us in a world of exhausted energy and mineral resources, chronic environmental disaster, gross overpopulation, and insufficient food. Instead, energy prices have fallen, food is plentiful if poorly distributed, birth rates are down, and pollution is increasingly under control. It also helps that the United States is not directly involved in a major war in which advanced technologies inflict massive damage on civilian populations and the countryside. Thus, the urgency of environmental and ecological issues that was apparent in the prior era has given way to a more bureaucratic, institutionalized, and embedded set of practices and ideals that see dealing with the undesirable side effects of new technology as a necessary but manageable aspect of its development.

THE NEW POLITICS OF TECHNOLOGY

These changes have many implications for the politics and conduct of technology policy. History shows that the primary actors in commercial technological change are industrial enterprises, and in the United States these are, overwhelmingly, private firms. Ordinarily, government scientists, engineers, and laboratories play at most a supporting role in commercial innovation, and successful government policies to stimulate private sector innovation are typically more indirect than direct.

Not all federal policy makers are comfortable with the idea that programs to stimulate technological development are more likely to succeed when they permit enterprises to respond in diverse and unpredictable ways to indirect incentives and opportunities for innovation. Some are committed to the view that government assistance should be accompanied by government participation in the decision-making process, and they would prefer to have government agencies play a substantial role in targeting assistance to specific technologies, such as those that promise to offer greater social benefits.

However, government decision processes, which emphasize fairness, openness, prudence, and accountability, are often incompatible with the timeframe, secrecy, risk-taking, and temporarily monopolistic character of private investments in technological change. These differences help explain why it has not been easy either to propose or to agree on an appropriate federal role in assisting industrial innovation.

Concurrent with the changing emphasis in technology policy is a changed pattern of business and government relationships. From the highly adversarial mode of the recent era of growth in government regulation of the side effects of technological applications, business and government relationships are evolving toward more cooperative or at least less hostile modes than prevailed a decade or so ago. This has caused strains on both sides, as each sector continues, wisely perhaps, to distrust the motives and practices of the other. For example, recent much-publicized industrial and financial practices, like hostile takeovers and insider securities trading, have undermined the emergent spirit of cooperation. And, some observers fear that stronger business and government ties will subvert the traditional role of government as watchdog over business excess.

Organized labor is of two minds regarding technological competitiveness. On the one hand, many workers and labor leaders recognize that the use of new technology is essential to employers' abilities to compete internationally, and many agree that the new factory technology demands changes in work rules and worker roles, if it is to be successful. At the same time, there is concern that widespread adoption of new technologies, especially of new process and manufacturing technologies, will lead to labor productivity increases without concomitant growth in output, with the result that fewer manufacturing jobs will be available. Furthermore, many remain concerned that the new manufacturing technologies require higher skills at the entry level, as well as from experienced workers, and that this will inhibit both entry and advancement for many workers who are not prepared to cope with the new demands.

Environmentalists also have a divided perspective on technological competitiveness. Like organized labor, they recognize the importance of remaining technologically competitive, as evidenced by their willingness to seek mutual accommodation with industrial interests in such matters as regulation of new, potentially toxic chemicals; cleanup of hazardous waste sites; and mandated home-appliance energy-efficiency standards. However, only a short time ago many industrialists attributed their competitiveness problems to the costs and barriers created by new environmental, health, and safety standards, so some environmentalists are also concerned lest the current enthusiasm for industrial competitiveness be turned against the environmental gains they have achieved.

Another aspect of the changed environment for technology policy is that the states have become powerful and independent participants in this arena.

They have established a plethora of programs including cooperative research and development (R&D) programs, industrial extension services at state universities, venture capital funds, and incubators for newly formed businesses. In aggregate, the states have budgeted several hundred million dollars for technology-based industrial development in recent years. In part, this more aggressive posture of the states, as compared with the federal government, reflects the long-standing role of the states as promoters of industrial and economic development. It also reflects the fact that many of the states are led by activist and pragmatic governors of both political parties, who are more inclined to seek a role in government in industrial development than is the Reagan administration, which seeks to limit this role as a matter of principle.

CONGRESSIONAL ACTION ON TECHNOLOGICAL COMPETITIVENESS

What kinds of policies has Congress adopted or examined to address the challenge of technological competitiveness? Where might it turn in the foreseeable future? My time frame for the past includes the two most recent Congresses (the 98th and 99th, running from January 1983 through January 1987). As for the future, I will venture only into the 100th Congress, running from January 1987 to January 1989.

Space will not permit a complete review of all of the legislative initiatives taken, or of all the legislation passed during these four years, so this is necessarily an overview. I focus on Congress, but it should be understood that many of the actions discussed have originated in the executive branch, or have had its support. Congressional problem solving is severely constrained without executive branch cooperation, or at least its acquiesence. It is rare for all of the members of Congress to agree on anything, so when I refer to Congress, I am usually talking about parts of Congress, and not always about a majority, but about vocal or well-organized parts.

Congress has been quite interested in promoting intersectoral cooperation in research and technology development. It has responded enthusiastically to the National Science Foundation's Engineering Research Centers program that began in 1985, which funds university-based interdisciplinary centers for basic research and education focused on important areas of industrial interest. These centers must have assurances of substantial industrial funding in order to quality for NSF funding. In 1984 Congress passed the National Cooperative Research Act, which facilitates joint research and development activities among firms in the same industry, and which provides special protections against certain kinds of government or private antitrust actions that might ensue from such cooperative research. Congress has also been a strong supporter of the Small Business Innovation Development Act of 1982, which requires federal agencies with substantial R&D budgets to set aside a percentage of their budgets to support small-firm development of new technologies in which the government has an interest. In essence, this Small Business Innovation Research (SBIR) program has put the federal government in the early-stage venture financing business in selected areas of technology.

Congress has also been interested in strengthening the human resource base for our technologically competitive society. It has increased funding for university research and graduate fellowship programs, and it has also been interested in enhancing the mathematical and scientific literacy of students and citizens at all levels, as well as in the specialized training of technicians. Congress has earmarked appropriations for new research facilities on a number of campuses, and, although these actions have often been controversial, they reflect a Congressional view that research, and technology derived from research programs, can play important roles in local economic development. Congress has done relatively little to upgrade university science facilities and equipment generally, although a number of proposals have been considered, and it has encouraged the Department of Defense in its establishment of a new University Research Initiative to fund university research equipment.

Some in Congress have raised questions about the continuing shift of federal F&D resources toward the defense sector and away from civilian purposes, because they are concerned that this may undermine the base of science and technology that is useful to industrial technology. In this context, however, it should be noted that a large fraction of the civilian R&D budget is directed toward civilian purposes that also have little to do directly with international economic competitiveness, such as health, space research, energy supply, and regulatory compliance. Only a small proportion of the federal science and technology budget has been directed toward civilian industrial needs at any time in the post–World War II era.

Strengthening the intellectual property system, especially patent protection for new inventions, has been another major congressional concern. Patents are intended to enhance the ability of inventors to reap rewards from their efforts by giving them a temporary monopoly over the use of their ideas. At the same time, publication of the details of inventions in patents is intended to inform other inventors about new product or process concepts that they may be able to use or extend if they make appropriate arrangements with holders of existing patents.[6] In recent years, Congress has taken several steps to transfer ownership of patents awarded for inventions made under government contracts to contractors, in hopes that this will spur the additional investments needed to transform inventions into commercial products. Similarly, in 1986 Congress gave the federal laboratories the authority to award title to patents deriving from cooperative research projects to the private individuals and corporations that participate in them. Congress also allowed government agencies to share royalties received for use of government-owned patents with the responsible government-employed inventors. The 98th Congress created a new class of intellectual property, semiconductor "mask works," and protected it from unauthorized use under the Semiconductor Chip Protection Act of 1984.

Late in the 99th Congress, time ran out for the consideration of a bill to give additional legal protections to owners of U.S. process patents that are used without permission by firms overseas to produce products that are then imported into this country. This bill was controversial, both because some U.S. importers of products such as generic pharmaceuticals opposed it and because it created new powers not now available to a class of U.S. patent holders that would extend beyond the intended target of imported products. Nevertheless, consideration of this bill is an indicator of the congressional interest in plugging holes in the patent system that are thought to impair the ability of U.S. firms to compete internationally. This bill is likely to be considered again in the 100th Congress.[7]

Closely related to the issue of patent protection is the issue of technology transfer, especially the transfer of technology from federal laboratories and contractors to industry. Although formal activities in this area date back to the State Technical Services Act of 1965, Congress gave renewed emphasis to the role of laboratories as providers of technology in the Stevenson-Wydler Technology Innovation Act of 1980, which requires all major federal agencies that conduct research and development to establish formal programs to transfer their findings and technology to the private sector. This act was amended and strengthened by the Federal Technology Transfer Act passed in late 1986. The effect of these laws is to give each major federal research laboratory the responsibility, in addition to and when compatible with its primary mission, to seek opportunities to transfer to industry, or to state and local governments, any potentially commercializable technology it develops.

"Technology transfer" has another connotation: the leakage of advanced technologies with military uses to potentially hostile powers. For some years, a debate has raged over the proper balance between keeping advanced technology secure for defense purposes, and encouraging exports of American high technology products to friendly nations for economic purposes. Torn between two high-priority objectives, Congress has wrestled repeatedly with this question in consideration of the Export Administration Act, the law that establishes the framework within which this balancing is carried out for specific technologies and specific transactions. The result has been what many, especially those in civilian industries, see as an overly cumbersome decision process involving elements of the Departments of Defense, Commerce, Treasury, and State. The debate turns not only on differences of view as to the nature of the military threats in the world, but also on different understandings of what is involved in effective "transfer" of technology, whether to another firm or to a hostile nation. It would appear that the question has not yet been completely settled, and the debate continues.

One other area to which Congress has devoted some attention, but without taking significant action, is the organization of the federal government to assist in technology-based industrial development and transfer. Proposals have been considered for various new organizations, including the creation of a Department of Science and Technology at the most ambitious, the establishment of an independent National Technology Foundation or Advanced Technology Foundation, the creation of a National Bureau of Standards and Industrial Competitiveness based in the existing National Bureau of Standards, and the redesignation of the National Science Foundation as a National Science and Engineering Foundation. None of these steps have been taken, although they have been considered from time to time by individual committees of Congress. No consensus has been reached on what such a structural change is intended to accomplish, or on what form of organization, if any, might be most useful and effective. The Reagan administration has not expressed significant support for any of these concepts, and in the absence of executive initiative it has generally proved difficult for Congress to engage in federal organization.

THE FUTURE

In view of the new attitudes toward technological competitiveness reflected in these congressional actions, where might we be going in the near future? First, I would argue that the shifts of the focus of technology policy from control to promotion and, within promotion, from specific national needs to technological competitiveness are large, important, and enduring. They are likely to survive the change in administration in 1989, regardless of the party of the new president, and they are likely to remain as the foundation of congressional interest and concern as well.

The shifts in attitudes toward technological change go beyond the United States; in fact, these shifts are repeated around the world. They appear not only in the United States, but also in both Western and Eastern Europe, in the Pacific rim countries including Japan, in the People's Republic of China, in India, and even in the Soviet Union. All are responding to the challenge of the new international competition that is based not on *exploiting* comparative advantage, but on *creating* it through new technology-based industry.

It would be a misinterpretation of current trends to assume that the commitments of citizens, governments, and industry to environmental quality have weakened as a result of the new focus on competitiveness. To be sure, a new spirit of accommodation and compromise exists in certain quarters, but the public remains determined to support efforts to clean up old environmental problems and prevent new ones, even if the costs exceed those now being paid. At the same time, many in industry realize that it is impossible to sustain rapid economic growth and industrial development without adequate attention to the problems they pose for health, safety, environmental quality, safe working conditions, and the character of work.

Congress can be expected to pay serious attention to issues of technology and industrial competitiveness in 1987. The passage of the Tax Reform Act of 1986 along with the continuing reasonable performance of the economy will give the committees concerned with economic affairs some breathing room to examine these questions, keeping in mind that their first priority is likely to be managing the budget deficit.

By mid-1987, the Science Policy Task Force of the House Committee on Science and Technology will have completed its inquiry, started in 1984, into the health of basic and applied research in government and the universities, and the Committee has indicated that it will turn its attention to the issue of technology policy in the 100th Congress. In addition, late in the 99th Congress, a new Congressional Competitiveness Caucus was established, whose members from both Houses and both parties are concerned to move international competitiveness higher on the national agenda in 1987.

The United States is trying to cope with a new set of international economic challenges from technically advanced industries overseas at the same time that its macroeconomic policies are putting domestic industries across the board under increased pressure from lower-cost foreign producers. Engineers and scientists hold out the promise that science and technology can improve the productivity of our industries and make them more competitive, and Congress is currently quite interested in helping them get the job done. There is some risk for the technical community in this promise, because the pressures of macroeconomic forces could make it impossible to succeed. The prudent course would be for scientists and engineers to turn themselves to the task of enhancing the technological basis for international competitiveness with great vigor, while seeking no great publicity for trying.

POSTSCRIPT

Shortly after this paper was drafted, the election of November 1986 resulted in the control of both Houses of Congress by the same party, and this may enhance the chances for agreement between the two Houses on policy issues related to technological competitiveness. Members of Congress of both parties seem agreed on the importance of competitiveness to the nation. President Reagan made competitiveness a major theme of his January 1987 State of the Union address. The competitiveness agenda is even more full than I anticipated in October 1986.

NOTES

This article is based on the GTE Foundation Lectureship on Science, Technology, and Social Progress, Lehigh University, Bethlehem, Pennsylvania, 9 October 1986.

The author wishes to thank Sarah Taylor, Irene Stith-Coleman, Peter Smit, Terrence Lisbeth, Steven Goldman, and Stephen Cutcliffe for helpful comments on an earlier draft. The views expressed herein are his own and not those of the Congressional Research Service.

1. Jefferson's Declaration of Independence spoke of inalienable rights, including life, liberty, and the pursuit of happiness, which in his view a person was not free to give up—i.e., to "alienate." See Garry Wills, *Inventing America: Jefferson's Declaration of Independence* (Garden City, N.Y.: Doubleday & Company, 1978), pp. 213–55.

2. Both the impetus to social control of technology and the faith in its social utility can be attributed to the belief in its great power.

3. Donella H. Meadows et al., *The Limits to Growth* (New York: Universe Books, 1972).

4. Barry Commoner, *The Closing Circle* (New York: Alfred P. Knopf, 1971).

5. W. Paddock and P. Paddock, *Famine 1975! America's Decision: Who Will Survive?* (Boston: Little, Brown & Company, 1967).

6. A patent does not give its holder a license to exploit his invention, but gives him the right to prevent others from using it unless they compensate him. The holder of a patent may be unable to exploit his own invention if this would require that he exploit others' patents.

7. A provision of the Omnibus Trade and Competitiveness Act, passed in August 1988, addressed this problem.

Technological Frontiers and Human Integrity

LANGDON WINNER

Long ago Archimedes is reported to have said, "Give me where to place my lever and I will move the world." If one considers the astonishing power made available by scientific technology in the late twentieth century, it appears that a number of Archimedean fulcra have been discovered at last. The world-altering possibilities presented by microelectronics, bio-technology, nuclear energy, and other rapidly evolving techniques would seem to provide an occasion to pause and take stock, to ask questions about where we are headed and why. Which criteria ought to guide our choices? Are there sources of wisdom about good and evil that might inform our policies? Are there limits to the kinds of development we will pursue?

There have been moments recently in which our society has seemed ready to begin this work of serious reappraisal: the moratorium on recombinant DNA research called by molecular biologists in 1974; the debate on limits to growth that continued throughout much of that same decade; the proposals for a nuclear weapons freeze of the early 1980s. But as promising as such developments seemed at first, all of them eventually lead to a similar result—a renewed commitment by leading figures in politics, science, engineering, business, and education to forge ahead with renewed speed and dedication along exactly the path that had somewhat earlier been noted as an object of concern. Thus, the biologists' moratorium was soon followed by a period of great enthusiasm for founding research laboratories and corporate enter-prises in biotechnology; the debate on limits to growth was resolved in favor of a renewed commitment to growth; the quest for a nuclear weapons freeze was eclipsed by a renewed drive to build not only the MX missile but to begin work on a multitrillion dollar system of space weaponry.

For anyone who studies the ethical and political dimensions of modern technology, recurring episodes like these offer an endless source of fascina-tion. Why does each pause to contemplate our situation in the face of obvious troubles and pressing human needs merely produce a new burst of enthusi-asm for the race ever onward? What is it that drives us so frantically in this direction?

One obvious explanation comes to mind. For more than a century, many of our society's most important social institutions have been structured to promote discovery and innovation. Research and development laboratories in business, government, and the university have as their explicit goal the production of new knowledge and technical breakthroughs. The success of corporate marketing strategies depends upon a rapid turnover of new products, model changes, and fashion contours. To be caught selling last year's model, even a perfectly good one, is taken to be a sign of weakness. The same is true of advancement within professional careers, including academic careers, which are predicated on a rapid turnover of shiny new intellectual products; scholars and scientists are rewarded for producing new knowledge rather than, for example, for seeking the wisdom needed to make better use of knowledge that already exists. This restless process is fueled, of course, by intense competition at home and abroad. If we do not come up with the next discovery, invention or idea, someone else—the Japanese, Soviets, or others—certainly will. A very broad, deep consensus maintains that the whole process is to the good, creating as it does an ever-expanding, ever-improving world of industrial wealth, consumer goods, health care, national security, and the like. Such is the version of the idea of progress that prevails in our time.

But there is ultimately something unsatisfactory in explanations of this kind. What they overlook is the deeply passionate attachment our society has to technological advance seen as an end in itself. To understand this commitment, one almost totally absent in human history before the nineteenth century, it helps to notice that it is not entirely rational. Rooted in colorful dreams, myths, and fantasies, expressed in recurring rituals that contain an enormous spiritual energy, the nation's involvement with technology is, among other things, clearly a matter of religious zeal.

Such feelings were prominently displayed during the week or two following the explosion of the space shuttle *Challenger*, 28 January 1986. Almost to a man and woman, the nation took part in a remarkable television-centered ritual of grief and rededication. That people felt the need to mourn the deaths of the seven astronauts is both understandable and laudable. But as one watched the spectacle of grief covered in exquisite detail virtually nonstop on all three networks, it was clear the nation was also experiencing something even deeper. Statements by politicians, television commentators, and people on the street expressed a uniform belief that whatever else one may say about it, the ultimate meaning of the tragedy was a need to rededicate ourselves to the space program and, indeed, to technological progress itself. Again and again one heard the uniform and unwavering assertion: we must not be deflected by tragedy; we must move ahead on the path blazed by these brave pioneers. Librarian of Congress Daniel Boorstin commented that they were explorers in the great tradition of Sir Francis Drake and Ferdinand Magellan.

President Reagan said repeatedly that the astronauts and their vehicle embodied the hopes and dreams of the entire nation.

In that climate of opinion it seemed impossible to ask: What are the purposes of the space shuttle? Whose interests does the project actually serve? Isn't its mission primarily a military one? Who stands to benefit from its development and who pays the costs? How much of the space program—for example, the selection of a school teacher for its crew—can be attributed to a NASA public relations campaign? To have raised such questions would have appeared sacrilegious. During the months that followed, journalists, scholars, congressional committees and a presidential commission revealed both the short- and long-term circumstances that lead to the disaster. But by then rituals of reaffirmation were complete.[1]

To explore the range of dreams, fantasies, and myths at the root of our attachment to technological advance is, of course, an enormous undertaking; the origins of this faith are both complex and deeply buried in our culture. Metaphysical and mythical ideas about linear time, human perfectibility, heroism, and the dream of conquering nature constitute the foundations for even the most ostensibly rational of modern projects. But there is one theme within the idea of progress whose powerful hold on Americans sets it aside for special scrutiny, an idea that locates prospects for human well-being on something called a *frontier*. It is that notion, its significance and consequences, that I want to discuss here.

AMERICAN JEREMIAD

One important tap root of distinctly American conceptions of progress can be found in the experience of the Puritans who migrated to America in the early seventeenth century. The New England Puritans believed that they were leaving behind a world hopelessly infected by sin and corruption and that they were moving to a New World in which the will of God would be realized. They had entered into a covenant with God and with each other to repent of sin and accept the Lord's grace. Aboard the flagship Arbella in the middle of the Atlantic en route to this promised land, John Winthrop summarized their mission. "The end is to improve our lives to do more service to the Lord, the comfort and increase of the body of Christ whereof we are members, that ourselves and posterity may be the better preserved from the common corruptions of this evil world, to serve the Lord and work out our salvation under the power and purity of His holy ordinances." The success of this mission into the wilderness would then serve as an example, a beacon of light to stimulate the reform everywhere else. "For we must consider that we shall be as a city upon a hill, the eyes of all people are upon us. So that if we shall deal falsely with our God in this work we have undertaken, and so cause

Him to withdraw His present help from us, we shall be made a story and a by-word through the world."[2]

The mission of the New England Puritans was ambitious, high-minded, and transcendentally hopeful. It rejected the medieval notion that human striving involved the sin of pride and was, therefore, ultimately doomed to failure. The Puritans believed that if a community of Christians could maintain a high enough level of devotion and discipline, it could create a lasting community worthy of God's blessing. At the same time, they were profoundly aware of the snares and pitfalls awaiting them at every turn. The lures of the material world and infirmities of their own carnal and spiritual natures conspired to produce the most hideous possibilities for sin against which they struggled with limited success. In the decades after they had established their settlements, however, it became clear that the extent of their backsliding and corruption was, especially by their own standards, considerable. Sermons preached by American Puritan clergymen again and again lament signs that the community had fallen away from the mission that justified its presence in the New World. In jeremiads that rang out from New England pulpits, the "visible saints" were excoriated for their failures and offered a vision of God's wrath that suggested impending doom.

Standard interpretations of New England Puritan experience, the writings of Perry Miller most notably, have stressed the sense of utter despair that fills the sermons of later generations of Puritan clergymen. But as historian Sacvan Bercovich has convincingly argued, these very same jeremiads contain a consistently optimistic theme. The faithful were told that although they had failed in their commitment to create a better world, they should take heart: their goal would be realized at some point in the future. As John Higginson advised his congregation in 1697, "if we look on the *dark side,* the *humane side* of this work, there is much of humane weakness, and inperfection; [but] if we look on the light side, the *divine side* of this work, we may see, that God hath established His covenant [to] be the God of his *people* and of their *seed* with them, and after them. And therefore *all that the Lord hath done for his people* in New England may stand as a monument, in relation to future times, of a fuller and better *reformation* than hath yet appeared, those times of greater light and holiness which are to come, when the Lord shall make Jerusalem a praise in the earth."[3]

In Bercovich's analysis, the lament in the New England jeremiad "was a strategy for prodding the community forward, in the belief that fact and ideal *would be made* to correspond."[4] "The essence of the sermon form that the first native-born American Puritans inherited from their fathers, and then 'developed, amplified and standardized,' is unshakable optimism. In explicit opposition to the traditional mode, it inverts the doctrine of vengeance into a promise of ultimate success, affirming to the world, and despite the world, the inviolability of the colonial cause."[5] Thus, the sense of gloom powerfully expressed at one moment is quickly resolved in the reassertion of the wonderful promise of accomplishments expected in the future. The rhetoric of this message was a powerful, durable form of discourse that helped shape the community's sense of purpose.

In succeeding generations the colonists moved from religious to increasingly secular pursuits. In particular they began to settle further and further into the wilderness. But the idea of entering a New World remained. No longer was it limited to New England alone; the promised land eventually became the whole of North America. Hence, the challenge of creating a new world was relocated on the American frontier. Bercovich argues that long after the Puritans had faded as a primary influence in American life, the rhetorical form of their jeremiad—lamentation over the gap between ideals and reality accompanied by a proclamation that the future would surely bring fulfillment of the dream—still continued to shape American beliefs. It entered the nation's political thought, its poetry, fiction, and its sense of history.

In this light, one figure of particular interest is the great historian of the American frontier, Frederick Jackson Turner. Writing in the late nineteenth and early twentieth centuries Turner reformulated the vision of America as a place of renewal, arguing that he was simply observing a pattern clearly evident in historical fact. "The existence of this land of opportunity has made America the goal of idealists from the days of the Pilgrim Fathers. With all the materialism of the pioneer movements, this idealistic conception of the vacant lands as an oportunity for a new order of things is unmistakably present."[6] In Turner's view the whole of American character depended upon the people's encounter with a vast continent. They had indeed left an Old World to enter a completely new one; that New World was the American West. From their experience of conquering the land grew the Americans' taste for democratic institutions, their sense of self-reliance, egalitarianism, and passion for freedom. Indeed, in his view, the development of American democracy was a reflection of a process in which individuals moved ever westward, deliberately removing all social and political restrictions upon their mobility.[7]

Surveying the situation at the close of the nineteenth century, however, Turner noticed an obvious fact. With the settlement of the West virtually complete, America's geographical frontier was at an end. Meanwhile industrialism was producing dangerous concentrations of wealth and social-class divisions that threatened the vitality of democracy. He wondered openly whether the great promise of freedom and democracy in the United States would be destroyed by corruptions similar to those that had long afflicted Europe. How would the country be able to sustain and continue what was strong and noble in the frontier spirit?

In an essay written in 1910, "Pioneer Ideas and the State University," Turner believed he had found an answer. The next great frontier would be found in the expanding horizons of scientific research. "New and beneficent discoveries in nature, new and beneficial processes and directions of the growth of society, substitutes for the vanishing material basis of pioneer democracy may be expected if the university pioneers are left free to seek the trail." He called upon the "chemist, physicist, biologist and engineer" to master "all of nature's forces in our complex modern society." In Turner's eyes, the great abundance once provided by readily available land and natural resources of America would give way to abundance generated by advances in science-based industry and agriculture. "The test tube and the microscope are needed rather than axe and rifle in this new ideal of conquest."[8]

My point here is not to stress the importance of Turner's thesis, but simply to note that he was reworking a theme deeply present in American consciousness. At about the same time, other writers were drawing similar conclusions. In 1901 the historian Charles Beard argued that the traditional strength of American democracy, the agrarian way of life, would be replaced by an industrialism equally favorable to democratic government. It was simply a matter of moving from one kind of promise to another, from one source of material plenitude to another one even better.[9] During the decades of the late nineteenth and early twentieth centuries there were a great many spokesmen who proclaimed that science and technology were to be the salvation of the America and the hope of mankind. At a time in which it was clear that the frontier and the hope it carried had to be somehow redefined, science was one promising contender, although there were others as well, including American power overseas and the reform of the industrial city. It remains an interesting question (but one I shall not address here) how in the late nineteenth and early twentieth centuries science was able to achieve such ideological prominence, a power over the imagination that has endured to this day.[10]

If one examines the moral and political dimensions of the ideas that have shaped American optimism for the past three and a half centuries, two features immediately stand out. American ideas about progress, ideas about moving from an Old World to a New World, are capable of seemingly endless transformation without losing their underlying message. When it appears that we have failed in the mission of building a City on the Hill to the glory of God, we simply place the realization of that off in the future, affirming that we are making progress toward it. When it becomes clear that the religious quest has faded as the primary goal, the mission is refocused on the spiritual challenges and material rewards involved in taming the wilderness, conquering the frontier. And when the geographical space that defined the frontier is used up, the vision is yet again reformulated, this time defining the frontier as one of scientific and technical advance feeding economic growth. This last turn is truly a fortunate one in this evolving system of beliefs. For there appears to be no limit to the frontiers presented by research and development—at last, a frontier stretching to infinity.

A second notable feature of this optimistic conviction is that it excludes any strong notion of responsibility. Because the fulfillment of the dream lies

in the future, neither the community nor any of its members can be blamed
for unfortunate conditions that prevail in the present. One's experience is
regarded as a continuing process of exploration, discovery, and spirited risk
taking. Nothing ventured, nothing gained. Defined in this way one's life
becomes a prolonged childhood. In the words of political scientist Michael
Rogin, "The 'great, powerful, enterprising and rapidly growing nation' (John
Quincy Adams) never had finally to face the consequences of growing up. It
could begin again on the frontier."[11] Within this mood of national self-
confidence, there is no need to take stock, no need to ask "Who are we?" or
"Where ought we to be going?" The answers are assumed. We are the people
moving outward, working on the cutting edge. Confronted with indications
that things are not what they should be, we point West, point "forward,"
point to things yet to come, "to boldly go where no man has gone before."[12]

FRONTIERS OF RESEARCH

Since Turner's discovery of the horizon provided by science and technology, the frontier metaphor has again and again been employed to guide
Americans as they struggle with important choices about research, development, and application. During the final days of World War II, for example, the
question arose as to what the nation's science policy ought to be after the
war. There were even some proposals that publicly supported science be
guided by a democratic process establishing specific social goals for research. The task of arguing a coherent position for the scientific community
was delegated to Vannevar Bush, an electrical engineer and physicist who
was president of the Carnegie Institute and director of the government's
Office of Scientific Research and Development. In a document submitted to
President Truman in July 1945, Bush argued that large amounts of federal
funds should be devoted to the support of science. But, he advised, the
agenda for research ought to be left strictly to the scientists themselves to
decide. The title of the report was, of course, *Science: The Endless Frontier,*
and its message draws heavily upon the classic metaphor. "The frontiers of
science must be thrown open so that all who have the ability to explore may
advance from the farthest position which anyone has attained."[13] In this case
the frontier image is explicitly offered as a way to discredit the idea that the
country might institute democratic controls of scientific and technical development. True pioneers, it would seem, need no guidelines from bureaucrats and politicians, only their continuing patronage.

In twentieth-century American popular literature about science and technology, the notion of frontier is far and away the most important metaphor
deployed in attempts to win public favor, even surpassing another favorite
notion—"revolution"—as a source of high-minded hyperbole. The notion
appears in hundreds of books and articles, proclaiming new frontiers in

everything from nuclear physics to extrasensory perception.[14] Hence, J. B. Rhine, parapsychologist at Duke University, adopted this rhetorical strategy in his book *The New Frontiers of the Mind* published in 1937. Rhine proclaims, "If from these further adventures we attain an evidential eminence from which still further frontiers of the mind are visible, who would prefer to have stood with Balboa on a peak in Darien for that initial sight of a new ocean or even on the bow of the Santa Maria for the first happy glimpse of the outlines of a new world!"[15]

All aboard. . . !

Appreciating the strength of the frontier idea within American history, one can inquire about the status of the idea in our own time. What are the frontiers that hold the greatest fascination now? By far the most prominent, of course, are those now labeled "high technology"—the challenging aspects of microelectronics, computer hardware and software, telecommunications, and biotechnology. This is where scientists and engineers find the challenging problems for research and development and where the populace is encouraged to invest its hopes.

One area of particular significance involves the developments in "high-tech" weaponry now being explored in the context of the Strategic Defense Initiative. In his March 1983 speech that announced the project, President Ronald Reagan invited "the scientific community in our country, those who gave us nuclear weapons, to turn their great talents now to the cause of mankind and world peace: to give us the means of rendering these nuclear weapons impotent and obsolete." Called "Star Wars" by Senator Edward Kennedy and other detractors (a term commonly used to describe the plan ever since), the plan proposes to build several layers of antiballistic missile defense using laser guns, particle beams, and other devices controlled by computers and, in some versions, powered by hydrogen bombs. For its most fervent supporters, however, the program has a different name, one that should not at this point surprise the reader in the least: "High Frontier," a name that echoes the title of space-colony advocate Gerard O'Neill's 1977 book.[16]

In his book, *We Must Defend America,* retired Lieutenant General Daniel O. Graham, former director of the Defense Intelligence Agency, argues enthusiastically for the "High Frontier," seeking to refute the many objections that have been raised against it. His specific arguments pro and con, although important in their own right, need not concern us here. What is interesting in this context are the images and metaphors General Graham uses to express the passion of his position. He writes of the fact that human aspirations were long hamstrung by being confined to the land. Then the Vikings and other explorers took to the seas in quest of wealth and military power. "At the beginning of the sixteenth century, after the epic voyages of men like Magellan and Columbus, human activity surged onto the high seas. Once again, the nations that mastered this frontier reaped enormous strategic rewards." He goes on to describe the recent extension of this impetus into space with the voyage of the astronauts to the moon and of unmanned satellites sailing forth on "the high seas of space." The old system of mutual assured destruction, he admits, has not worked; building these costly weapons systems in the name of national security has made us less rather than more secure. But the benefits of space technology will, he believes, go far beyond the need for defense. Space technologies, including those of the "High Frontier" project, "can eventually create the means to bring back to Earth the minerals and the inexhaustible solar energy available in space. By doing so we can confound the gloomy predictions of diminishing energy and material resources on Earth. This will not only enhance the prosperity of the advanced, industrialized nations of our Free World, but will also provide solutions to many hitherto intractable problems of developing countries."[17]

General Graham hastens to add that building a spaceborn defense does not mean that our nuclear retaliatory capabilities can be abandoned or neglected. He is also careful to stress that we must still "build and maintain strong conventional capabilities." But he concludes with a flourish, "We Americans have always been successful on the frontiers; we will be successful on the new high frontier of space. We need only be as bold and resourceful as our forefathers."[18]

One cannot, it seems to me, understand the Strategic Defense Initiative without understanding that it expresses American obsessions previously characterized by the likes of Daniel Boone and Davy Crockett and General George Custer. As the speeches of Ronald Reagan on this subject have made abundantly clear, the President sees space defense as the promise of a world redeemed. In effect, his call for lasers in space is a renewed vision of Winthrop's City on a Hill.

The rhetorical form of frontier discourse, as we have seen, easily absorbs or deflects criticism. In this case, many have argued against the proposed space weapons development plans on the grounds that they are so complicated and unworkable that they could not possibly do what they are supposed to. Such criticisms are greeted with the traditionally American response. The fulfillment of the dream lies in the future. What we need now is research. In this context *research* becomes the name of the epistemological quest that invests human inquiry in ever-expanding frontiers. But the variety of *search* that might accompany these efforts—a dedicated search to clarify the ends that orient our quest for knowledge and power—is notably missing from the project.

THE QUESTION OF INTEGRITY

Now that our society has explored the seas, conquered the land, pushed forward the boundaries of science, and launched forth into outer space, what frontiers remain? There is one avenue on our society's agenda that has an indelible fascination, one likely to prove even more significant than space as a horizon for research, development, and commercial exploitation. It is a frontier now at issue in a wide variety of research fields, including neurophysiology, biomedical technology, computer science, robotics, and molecular biology. What can be seen in such fields in the late twentieth century is the boundary between human beings and artificial systems targeted as a barrier to be overcome. This is truly the newest frontier, one even more novel and challenging than space travel, one that we need to ponder seriously before its conquest becomes an accomplished fact.

Inquiries along this boundary move in one or the other of two fundamental directions: inside out and outside in. The first of these, the inside out, involves attempts to master what were in earlier times exclusively human faculties and to transfer them to artificial devices of various kinds. In manufacturing, for example, this is a matter of replicating skilled or unskilled motions that would otherwise have been done by human workers. Such efforts have an even broader scope in the computer science of artificial intelligence where there are now enthusiastic attempts to reproduce or mirror such faculties as sight, speech, reasoning, and advising through the use of very high speed computers, sophisticated programming, and associated techniques. The idea upon which such projects are based holds that if one can identify the laws that govern a domain of cognition or behavior, one can then reduce these laws to a set of rules that can in turn serve as a basic program that can be executed within any suitable physical medium—a computer, for example. The fact that human beings exhibit intelligence is taken to be merely a sign that one such system embodies the crucial laws and rules but that scientists and engineers can develop artificial systems that embody them just as well. Many researchers in artificial intelligence believe that the extremely fast information processing made available by new generations of computers will eventually surpass human capabilities in important respects.[19]

The second path of research along this frontier, what I am calling the outside in, involves the search for increasingly intimate means of communication between humans and technological devices in ways that create a reciprocal functioning of natural and artificial organs. This is what is at stake in work on the artificial heart and other high-technology prosthetic devices, for example. There seems to be no end to the range of transplants and implants that modern medicine might devise to keep the human organism alive and functioning.

Microminiaturization in electronics provides many illustrations of the concerted push to dissolve the boundary between humans and technology. In California last year, a computer firm began marketing a product it calls "Mind Master." This device enables users of personal computers to communicate with their machines not through the use of a keyboard, not through the use of a joy stick or mouse, but through a galvanic skin sensor. The galvanic skin response is the process used in lie detectors. "Mind Master" uses this principle to move objects in ordinary video games. All one has to do is place one's fingers on a tiny box and then think or emote the movement of the jet airplane on the screen. That may seem a bizarre example. But in point of fact, the Department of Defense now spends hundreds of millions of research dollars in its Strategic Computing project to solve similar "problems." There is, for military planners, an increasingly serious bottleneck that involves the interface of humans and machines in high-technology weapons systems. Flying today's fighter planes, for example, places such great demands on the pilot—so much information has to be processed in such a short period of time under such physically stressing conditions—that human sensory and analytical capacities have to be augmented. Can one guarantee that the humans and machines are able to communicate? That gives rise to the search for ingenious new devices to manage the interface. The Air Force would like to invent an automated copilot that would communicate with the pilot through an intricate human technology symbiosis.[20]

Far and away the most significant research field in outside-in technology is that of genetic engineering. At present our abilities here remain somewhat primitive and our projects fairly modest. Scientists are working to develop new strains of edible plants that can survive under inhospitable conditions, and growth hormones and other pharmaceutical products to be used in health care. Genetic therapies are now employed in limited areas of medical treatment as a way to cure patients with previously intractable illnesses. But the likely agenda of research and development in molecular biology points to a time perhaps not too far distant when it will be possible to alter significantly the genetic composition of plants and animal life. At that juncture the power will exist to control evolution, including human evolution. The question may then become something like: What kind of human being would you like to engineer? Which traits would you select? Should the next generation be taller, smarter, more beautiful? Scientists have already grown genetically engineered mice that reach twice their normal size. Genes transplanted from fire flies have enabled tobacco plants to glow in the dark. Hence, would you care to order a ten-foot-tall basketball center? And will that be with or without bioluminescence?[21]

The list of research projects that confront the boundary between inside and outside could be extended much further. Although it is too early to anticipate exactly where such developments will lead, their general thrust points to a distinct possibility: that what has traditionally been called "humanity" on the one hand and "technology" on the other could eventually merge. An idea that Samuel Butler's *Erewhon* employs as a way to parody Darwinian theory—the notion that machines emerge as living organisms capable of evolution—is now taken as a serious proposition by some observers of contemporary advances in computer hardware and software. As one writer explains, "We see that the biological world will come to be characterized by three great classes of life—animal, plant, and machine. At the same time there will be certain independent groups; we have already met the protista (between animals and plants), and the progressive insertion of artificial components in animals, including man, which will yield new creatures mid-way between animals and machines."[22]

Indeed, there are some, perhaps even a great many, who would welcome this merger as a wonderful step forward. Computer scientist Marvin Minsky and science-fiction writer Arthur C. Clarke are at one in the opinion that the human species is best seen as a kind of transitional stage to higher, largely fabricated forms of existence. Minsky, who occasionally describes humans as "meat machines," forecasts the following symbiosis of humans and machines extended, a creature with unlimited spatial extension. "You don a comfortable jacket lined with sensors and musclelike motors. Each motion of your arm, hand, and fingers is reproduced at another place by mobile, mechanical hands. Light, dexterous and strong, these hands have their own sensors through which you see and feel what is happening. Using this instrument, you can 'work' in another city, in another country or on another planet. Your remote presence possesses the strength of a giant or the delicacy of a surgeon."[23] O brave new world, that has such people in it!

Such speculative projects and the actual inquiries that mirror them raise the question of how far we wish to travel into this new frontier, how thoroughly we want to settle it. These possibilities, regardless of how soon they arrive as feasible projects, raise most profoundly the question of human integrity, a question seldom at issue in human history until now because the meaning of the distinctively "human" has never faced such a momentous potential for alteration.

To say that humans have integrity suggests there is a wholeness, completeness, soundness, and authenticity in their being. Having evolved through millions of years they are well equipped in their present conditions of living. In this light, any techniques that would modify basic structures of the human organism ought to be regarded with extreme caution. Similarly, any techniques that intrude upon realms of thought and activity that have long been distinctly human pursuits ought to be subjected to the most severe scrutiny before any practical adoption. This does not mean we should not try to heal the sick, ameliorate the condition of the handicapped, or ease the burdens of human toil. Certainly we ought to investigate such means. But it is possible (or even likely) that in our Baconian understanding of such matters, our definitions of disease, infirmity, handicap, and toil will begin to expand in a manner that will identify as defects certain features that are simply part of basic human makeup, human faculties and, indeed, the human condition. Seen from a vantage point that holds scientific and technical advance to be a self-evident good, there is nothing that ought be sheltered from "improvement."

It is true that the works of human beings have already powerfully altered the course of our own evolution and of other species as well. Changes in nutrition, hygiene, shelter, medicine, and warfare, among other accomplishments, have modified the circumstances under which one generation passes its genetic content on to the next. But all previous changes are modest compared with the prospect that now faces us: approaching the boundary between nature and artifice as an elaborate set of engineering projects. Will we see ourselves as creatures with integrity, creatures who ought to be respected for what they are in their wholeness and coherence? Or will we see ourselves as bundles of discrete, separable features, each of which is ready to be renovated and reorganized? Whether we choose to face them or not, these questions are now implied by the current and future course of scientific research and technological development.

There is a common and convenient answer to this unsettling prospect. Of course, we would know where to stop. Humans would never do anything to undermine their own standing as a species, never willingly relinquish their supremacy over our own artifacts. But what is new and ironic about the situation we now face is that at last *we are both the pioneers and the frontier.* If the past is any guide, that is an extremely precarious position in which to find oneself. The American Indians, for example, found themselves in a similar predicament, living on what European settlers saw as frontier that was rightfully theirs. The Indians were, of course, swindled by bogus treaties, removed from their homelands, and brutally slaughtered as the nation expanded across the continent, a movement undertaken with great enthusiasm and little regret. Our civilization did not respect the integrity of Indian culture or the integrity of the Indians as living human beings. This was not, it pays to recall, a small aberration in the story of the American people. It was quite clearly a manifestation of the mission that brought settlers to this continent. Those who stood in the way of our cultural destiny eventually found themselves in the sights of a gun.

As the contemporary pioneering spirit approaches the fundamental boundaries of the human organism as such, will it exhibit similar aggressiveness and temerity? Or will it recognize limits upon what it is prepared to do? In modern history the need to recognize self-conscious ethical limits upon human activity is far from a popular notion. During the past two centuries there has arisen a remarkably strong practical and spiritual consensus that limits are simply onerous barriers that need to be surmounted. Those who master the tasks of overstepping limits receive tremendous material and psychological rewards. For that reason the basic inclinations of science, business, engineering, medicine, and advertising are geared to the continuing search for and celebration of "breakthroughs." Our culture contains a powerful momentum that makes any attempt to discuss limits seem nothing more than idle fantasy.

In the history of Old West, new settlements tended to precede the coming of law and order. Some of the most impetuous of the citizenry loaded their wagons and headed toward the new territories, hoping to escape, at least temporarily, the very laws and limitations that characterize civilized society. Frontier towns were notorious not only for their excitement and creativity, but also for occasional shootings and lynchings. The frontier village in our future is perhaps the human gene pool itself and it may be years before the new marshall arrives from Dodge City.

There is, of course, a middle ground of deliberation, occupied by scholars and citizens groups interested in the critical study of social ends and institutional forms involved in the practice of science and technology. Such criticism often includes the attention to alternative mixes of human ends and technical means. As is evident in the work of Jeremy Rifkin's Foundation on Economic Trends, such efforts sometimes have been able to pose the question of limits in practical, politically meaningful ways. Through law suits, lobbying and a campaign of public education, Rifkin and his colleagues have fought the introduction of genetically engineered commercial products marketed by rapidly growing biotechnology firms. While the specific aims of these attempts are important in their own right, even more significant is the way in which they seek to clarify the larger consequences of crossing the line that has heretofore separated natural and technological entities.[24]

But the critical understanding of science and technology, both in its scholarly and political dimensions, is not likely to gain widespread appeal. For its ideas and proposals typically appear to outsiders as merely "negative." As we have seen, an enduring requirement of American public discourse is that one's message ultimately appears positive and upbeat. Even if one believes that society faces severe ecological, social, or economic problems, even if one's analysis indicates that troubles lie ahead, the only truly acceptable strategy is to present oneself as a person who has found the silver lining, the ray of hope, the newest avenue of progress. In discussions about science and technology policy, this is an obsessive turn. Thus, for example, as he surveys the condition of U.S. industrial production in the 1980s, political economist Robert Reich warns of serious economic decline "marked by growing unemployment, mounting business failures, and falling productivity." His book, however, carries the inspiring title *The Next American Frontier* and goes on to describe a brighter day ahead as America adapts to a technologically dynamic, newly prosperous, socially well-managed, politically progressive future. "Adaptation," Reich explains, "is America's challenge. It is America's next frontier."[25]

The pattern here is familiar—the basic message of the American jeremiad. Although we recognize signs of failure along the present path, we are counseled to view them as good reasons to work all the more diligently in much the same direction the community has followed all along. At this point, however, perhaps a different kind of understanding is required. When our science produces weapons that threaten imminent extinction, when our economic and educational institutions define improvement as private rather than common wealth, when our standards of productivity suggest improvements in machines and the discarding of people, when the transfer of our science and technology to other nations fails to bring freedom, prosperity, and peace but produces tyranny, poverty, and conflict instead—perhaps it is time to admit there may be something fundamentally wrong with the whole program.

How we might escape the pattern of obsessive reaffirmation is a matter that merits serious attention. It will not be an easy task. One avenue that lies open to us is to find ways of talking about human well-being in a language not prejudiced and debilitated by worn-out concepts of frontiers and progress— and, for that matter, the concept of a New World on the horizon. Such notions have by now become a hindrance to any reasonable discussion of what it means to enhance the condition of human beings in our time. As an alternative we might talk about what a just society would look like, which specific ends it would seek to achieve, and which means, including forms of science and technology, are suited to it. Taking a concern for justice as a standpoint, we might explore the choices before us unprejudiced by the manic desire to advance for the sake of advancing. Similarly, if we want to talk about the good life and what it means to improve our ways of living together—always a rewarding topic—then let us take that up directly, rather than rely on threadbare, hand-me-down, shoot-em-up mythologies. If you find yourself playing a role in a never-ending John Wayne movie, is it not time to rewrite the script?

Perhaps we would profit as well by reconsidering why Western thought as exemplified by the New England Puritans rebelled against the medieval conviction that the very worst of the seven deadly sins was the sin of pride, the sin of believing that humans can do better than the Creator.[26] Clearly, that rebellion was justified in important respects; it opened many doors to new knowledge and, yes, to a general improvement in the material conditions potentially available to the mass of humankind. But in many ways that metaphysical insurrection now threatens to overwhelm even the rebel himself. An aphorism of Franz Kafka poignantly summarizes the situation of human integrity in an age of scientific technology: "He found the Archimedean point, but he used it against himself; it seems that he was permitted to find it only under this condition."[27]

NOTES

1. For a discussion of the role of the mass media before and after the explosion of the *Challenger*, see *Media Coverage of the Shuttle Disaster: A Critical Look*, A Panel Discussion Held at the AAAS Annual Meeting, Philadelphia, Pennsylvania, 26 May 1986 (Washington, D.C.: American Association for the Advancement of Science,

1986). As reporter Laurie Garrett notes, "I think there has always been a feeling in the NASA press office, no matter which press office you are talking about, that to be really critical of the space shuttle program was to be non-patriotic; to somehow question a basic Americanism; that to really ask the hard-boiled questions was not fair ball" (p. 12). For the official postmortem, see *Report to the President by the Presidential Commission on the Space Shuttle Challenger Accident* (Washington, D.C., 6 June 1986).

2. John Winthrop, "A Model of Christian Charity," in *The American Puritans, Their Prose and Poetry,* ed. Perry Miller (Garden City, N.Y.: Doubleday, 1956), pp. 82–83.

3. Quoted in Sacvan Bercovitch, *The American Jeremiad* (Madison: University of Wisconsin Press, 1978), pp. 96–97.

4. Ibid., p. 61.

5. Ibid., pp. 6–7.

6. Frederick Jackson Turner, *The Frontier in American History* (New York: Holt, Rinehart and Winston, 1920 and 1947), p. 261.

7. For a sample of writings that debate the significance of the frontier in American history see Ray Allen Billington, ed., *The Frontier Thesis: Valid Interpretation of American History?* (Huntington, N.Y.: R.E. Krieger, 1977); Richard Hofstader, *The Progressive Historians: Turner, Beard, Parrington* (New York: Knopf, 1968); David W. Noble, *Historians Against History; The Frontier Thesis and the National Covenant in American Historical Writing Since 1830* (Minneapolis: University of Minnesota Press, 1965); Richard Slotkin, *Regeneration through Violence: The Mythology of the American Frontier, 1600–1860* (Middletown, Conn." Wesleyan University Press, 1973); Richard Slotkin, *The Fatal Environment: The Myth of the Frontier in the Age of Industrialization, 1800–1890* (New York: Atheneum, 1985).

8. *The Frontier in American History,* p. 287 and 284.

9. Charles A. Beard, *The Industrial Revolution* (London: S. Sonnenschein, 1901). See also David W. Noble, et al. *The End of American History: Democracy, Capitalism and the Metaphor of Two Worlds in Anglo-American Historical Writings, 1880–1980* (Minneapolis, University of Minnesota Press, 1985).

10. For a discussion of this topic see Ronald Tobey's *The American Ideology of National Science, 1919–1830* (Pittsburgh: University of Pittsburgh Press, 1971).

11. Michael Paul Rogin, *Fathers and Children: Andrew Jackson and the Subjugation of the American Indian* (New York: Knopf, 1975).

12. The phrase is, of course, from the spoken introduction to each program in the "Star Trek" television series. Indeed, the mission of the Starship Enterprise, traveling from galaxy to galaxy in search of new worlds, is a colorful extension of the American frontier myth. It is interesting to note, however, that one fundamental norm limits the kind of involvement that Captain James Kirk and his companions may have with alien creatures: they must do nothing to alter the way of life of the beings they encounter.

13. Vannevar Bush, *Science: The Endless Frontier* (Washington, D.C.: U.S. Government Printing Office, 1945) p. 181. See also Ken Hechler, *Toward the Endless Frontier: History of the Committee on Science and Technology, 1959–79* (Washington, D.C.: U.S. House of Representatives, for sale by the Superintendent of Documents, U.S. Government Printing Office, 1980).

14. Some representative writings in this vein are The Atomic Industrial Forum, Stanford Research Institute, Stanford University, *Atomic Energy, the New Industrial Frontier; Proceedings of a Meeting for Members and Guests, April 4 and 5, 1955, San Francisco, California* (New York: The Atomic Industrial Forum, 1955); Yvonne Baskin, *The Gene Doctors: Medical Genetics at the Frontier* (New York: W. Morrow, 1984); Eric Burgess, *Frontier to Space* (New York: Macmillan, 1956; Milton Stanley

Livingston, *Particle Physics: The High-Energy Frontier* (New York: McGraw-Hill, 1968); Robert W. Lockerby, *Space Colonies: The Next Frontier* (Monticello, Ill.: Vance Bibliographies, 1980); United States Congress, Office of Technology Assessment, *Genetic Technology: A New Frontier* (Boulder, Colo.: Westview Press, 1982); Wernher Von Braun, *Space Frontier* (New York: Holt, Rinehart and Winston, 1967).

15. J. B. Rhine, *New Frontiers of the Mind* (New York: Farrar & Rinehart, 1937), p. 274.

16. Gerard K. O'Neill, *The High Frontier: Human Colonies in Space* (New York: William Morrow, 1977.)

17. Daniel O. Graham, *We Must Defend America and Put an End to MADness* (Chicago, Ill: Regnery Gateway, 1983), pp. 109–111.

18. *We Must Defend America,* p. 114.

19. See for example Ian Benson, ed., *Intelligent Machinery Theory and Practice* (New York: Cambridge University Press, 1986); Donald Michie and Rory Johnston, *The Knowledge Machine: Artificial Intelligence and the Future of Man* (New York: W. Morrow, 1985). For a sympathetic but critical view of work in this field, see Terry Winograd, *Understanding Computers and Cognition: A New Foundation for Design* (Norwood, N.J.: Ablex Publishing, 1986).

20. See for example *Strategic Computing: New-Generation Computing Technology: A Strategic Plan for Its Development and Application to Critical Problems in Defense,* Defense Advanced Research Projects Agency, 28 October 1983.

21. For a discussion of recent developments in biotechnology, see Martin Kenney, *Biotechnology: The University-Industrial Complex* (New Haven: Yale University Press, 1986), and Marc Lappe, *Broken Code: The Exploitation of DNA* (San Francisco: Sierra Club Books, 1984).

22. Geoff Simons, *The Biology of Computer Life, Survival, Emotion and Free Will* (Boston: Birkhauser, 1985), p. xii. See also Simon's *Are Computers Alive? Evolution and New Life Forms* (Boston: Birkhauser, 1983), and Samuel Butler, *Erewhon; Erewhon Revisited* (New York: Dutton, 1965).

23. Marvin Minsky, "Telepresence," *Omni* 2 (June 1980): 45.

24. See Jeremy Rifkin, *Declaration of a Heretic* (Boston: Routledge & Kegan Paul, 1985).

25. Robert B. Reich, *The Next American Frontier* (New York: Times Books, 1983), pp. 3 and 21.

26. For a discussion of the sin of pride in medieval thought see Morton Wilfred Bloomfield, *The Seven Deadly Sins* (East Lansing: Michigan State College Press, 1952).

27. Quoted in Hannah Arendt, *The Human Condition* (Chicago: The University of Chicago Press, 1958), p. 248.

Automation Madness, or the Unautomatic History of Automation

DAVID F. NOBLE

> Strange business, this crusading spirit of the managers and engineers, the idea of designing and manufacturing and distributing being sort of a holy war; all that folklore was cooked up by public relations and advertising men hired by managers and engineers to make big business popular in the old days, which it certainly wasn't in the beginning. Now, the engineers and managers believe with all their hearts the glorious things their forbears hired people to say about them. Yesterday's snow job becomes today's sermon.
>
> —Kurt Vonnegut, *Player Piano*

Today's sermon is a printout on the glories of computerized automation, of progress without people. And when the engineers and managers preach this gospel from electronic pulpits everywhere, they now truly believe their own snow job, and so too do we all. Together, we seek salvation in their scientific fantasies, calculate our chances with their egoistic economics, and accept their vision of the inevitable future as our destiny too.

Today our survival demands that we take another look at our ideas about progress and confront the sobering realities that these ideas allow us to ignore. For only then can we begin to challenge the self-annointed apostles of automation whose social irresponsibility is matched only by their madness. Put simply, we have come to believe in an automatic future, one driven and directed by an autonomous technological advance (technological progress) and leading inescapably to the best of all possible worlds (social progress). The first, we suppose, proceeds automatically and guarantees the second. Let us look at each in turn.

AUTOMATIC TECHNOLOGICAL PROGRESS

As a result both of ignorance and incessant inculcation by our established institutions, we have all come to hold a rather simple, though deceptively straightforward, view of how technology develops. In essence, this view is

Darwinian. We believe that the process of technological development is very much like biological evolution of the species through natural selection. Just as the earth's creatures evolve according to the anonymous and automatic logic of survival of the fittest, whereby only those forms best adjusted to the rigors of nature survive, so, too, do the myriad technological possibilities generated by human imagination and ingenuity pass through a competitive and thorough-going process of elimination, which guarantees that only those best suited to human purposes survive—as it were, naturally and automatically.

Because this view is deeply ingrained as a habit of thought, we rarely if ever think about it. But when we do, our half-conscious ideas look something like this: we assume that our technologies pass through two successive filters or screens, which automatically weed out the unsatisfactory contributions and allow only the best to emerge. The first, a technical or scientific screen, vaguely comprises the work of scientists and engineers who, with their dedication to rationality and efficiency, methodically subject all technological possibilities to careful and objective scrutiny and select only the best solution to any given problem. It remains something of a mystery exactly how this is accomplished, but we rest assured that it is.

The second screen is an economic filter, composed of two equally vague mechanisms. Upon successfully completing the technical test, the selected technologies are subjected to the no-nonsense, cost accounting, profit-maximizing evaluation of hard-headed, practical businessmen who seek only the most economically viable technologies from among those deemed technically superior. The "real-world" savvy of the businessman, we assume, corrects for the excesses of the less practical scientists and engineers. Finally, because even businessmen and their managers can make mistakes in judgment, we rely ultimately upon the fail-safe test administered automatically by the anonymous operation of the self-regulating market, which allows only the most economically astute businessmen to survive the rigors of competition and, with them, only the best technologies.

Thus, when we see a technology in regular and widespread industrial use, we confidently assume that it represents the best history had to offer, because it survived the successive tests of this process of natural selection. In this way we routinely dignify the present array of technology as the highest expression to date of so-called technological progress and, as such, we accept it as inevitable, a fact of life—beyond the realm not only of politics, but even of thought and discussion.

But if we take a more careful and critical look at this seemingly inevitable process of technical development, we recognize at once that it is not really automatic at all but political—something people plan for and struggle over. That is, we see that it is not some abstractly rational enterprise with an internal logic all its own, but rather a human effort that reflects at every turn the relations of power in society. This is the case for both the technical or scientific "screen" and the economic "screen" alike, as we will see.

Behind the Technical Screen

When we look past the veil of mystery that enshrouds the work of technical people, we find that their activities reflect their relation to power at every point. Their link with power gives them power—it entitles them to practice their trade in the first place, to learn, to explore, to invent; it emboldens their imagination; and it provides them the wherewithal to put their grand designs into practice. In short, it is the support of those in power that affords technical people the luxury to dream, to dream expansively and to make their dreams come true, by imposing them on others. Although most scientists and engineers would admit to their dependence upon those with power, few would concede that this relationship actually influences the way they think about things. They would insist, rather, that they are guided in their work by technical considerations above all else, and that this is what makes their calling rational and thus compelling. Moreover, having worked with and taught technical people, I know that few engineers are deliberately out to destroy jobs or unions or to harm people in any way. Although, of course, in practice they must satisfy the requirements of their boss, their client, or their customer, ultimately they aim only to do the best work for the good of society. Yet, consistently, they turn out solutions that are good for the people in power (management) but often disastrous for the rest of us (workers). Can this be explained?

For one thing, few technical people have any contact with workers. In their education and their professional careers, they typically communicate only with management. Not surprisingly, they tend to view the world pretty much as management does whether they know it or not. Although they are taught and usually believe that this view is objective, it is, in reality, the view from the top, the perspective of those with power.

For seven years I investigated the history of automated machine tools. Much of the pioneering design and development work took place at MIT and I spent many months pouring over the vast collection of documents from the ten-year project. I discovered that the engineers involved in creating this self-professed revolution in metalworking manufacturing had been in constant contact with industrial managers and industry officers who sponsored and monitored the project. Yet I found not a single piece of paper indicating that there had been contact with any of the many thousands of men and women who work as machinists in the metalworking industry—those most knowledgeable about metalcutting and most directly affected by the technical changes under development. The engineering effort was essentially a management effort and the resulting technology reflected this limited perspective—the world view of those in power.

By considering only those solutions compatible with power, engineers are guided, often unconsciously, by their own desires for recognition and power. Let me illustrate. Suppose I announced to an audience of technical people that I had developed an ingenious new system that could produce some indgit in half the time it takes conventionally, included the latest state-of-the-art components, and had been fully tested. The only thing the audience had to do was follow my instructions—do exactly what I told them for as long as I said. That is, a central operating feature of the system's design was that it gave me complete control over everyone else's activities. Even those who shared my enthusiasm for this system would resist giving me the right to make all of the decisions. If I pressed too hard, perhaps they would show me the exit. Yet, if I were to take that exact same design to Lee Iacocca or Henry Ford, or any top manager in industry, chances are they would consider me a genius, buy the system, and hire me to implement it. What exactly is the difference between the two situations such that, with the same invention, in the first case I might be ridiculed but in the second hailed as brilliant. The difference would be the relations of power. In the first instance, I do not have the power to get the audience to follow my instructions so my design can be dismissed. In the second case, however, the executive knows that he could compel his employees to do as I say, and so the same design is considered not only viable but a breakthrough.

To push this example a little further, suppose that the audience, instead of dismissing me and my design, succeeded in engaging me in a serious debate. The result was now a compromise design that was just as effective but required audience participation—a democratic design, so to speak. Now if I took this improved (and more challenging) design to the executive, he would be the one to dismiss it as absurd: What, a system that gives workers the same decision-making power as the manager? Nonsense. What are you some kind of radical? The point is this: that the viability of a design is not simply a technical or even economic evaluation but rather a political one. A technology is deemed viable if it conforms to the existing relations of power.

Engineers learn early on that in our society, the authoritarian pattern predominates in all institutions and workplaces. So when an engineer begins to design a top-down technical system, he reasonably assumes from the outset that the social power of management will be available to make his system function. Such authoritarian systems are also simpler to design than more democratic ones, because they entail fewer independent variables, and this also makes them more appealing to designers. Finally, authoritarian systems satisfy the engineer's own will to control, and offer the engineer a powerful place in the scheme of things. Thus, for all these reasons, new technical systems are conceived from the outset as authoritarian ones, perfectly suited for today's world. With little forethought and no malice to speak of, engineers routinely draw up designs and construct systems that concretely reinforce the power of those they serve. In the process, their own interests, ambitions, and compulsions become intertwined with and indistinguishable from those of their patrons. Never are all possibilities entertained and soberly evaluated, as the Darwinian idea of technological progress suggests; only those that are compatible with the authoritarian position and the disposition of those with the power to choose.

When I studied the history of industrial automation, I found that, although technical and economic considerations were always important, they were rarely decisive to the ultimate design and deployment. Behind the technical and economic rhetoric of justification I consistently found other impulses: a management obsession with control; a military emphasis upon command and performance; and enthusiasms and compulsions that blindly fostered the drive for automaticity.

CONTROL: AN OBVIOUS OBSESSION

Many academic studies today purport to describe and explain the advance of industrial automation, but few ever even mention a major impulse behind it: management's obsession with and struggle for control over workers. However much this control might be justified in the name of economic efficiency, with the self-serving claim, belied by nearly every sociological study of work, that centralized management authority is the key to productivity, the truth of the matter is that control is less a means to other ends than an end in itself. Above all else, management strives to remain management, whatever the costs. To this end, they consistently solicit and welcome technologies that promise to enhance their power and minimize challenge to it, by enabling them to discipline, de-skill (in order to reduce worker power as well as pay), and displace potentially recalcitrant workers. Perhaps more than any other single factor, this explains the historical trend toward capital-intensive production methods and ever more automatic machinery, which have typically been designed with such purposes in mind. This is an old story, really, one that was perhaps best told by Andrew Ure, one of the earliest apostles of industrial automation, in the 1830s, back at the dawn of the industrial revolution:

> In the factories for spinning coarse yarn . . . the mule-spinners [skilled workers] have abused their powers beyond endurance, domineering in the most arrogant manner . . . over their masters. High wages, instead of leading to thankfulness of temper and improvement of mind, have, in too many cases, cherished pride and supplied funds for supporting refractory spirits in strikes, wantonly inflicted upon one set of mill-owners after another. . . . During a disastrous turmoil of [this] kind . . . several of the capitalists . . . had recourse to the celebrated machinists . . . of Manchester, requesting them to direct [their] inventive talents . . . to the construction of a self-acting mule. Under assurance of the most liberal

encouragement in the adoption of his inventions, Mr. Roberts . . . suspended his professional pursuits as an engineer, and set his fertile services to construct a spinning automaton. . . . Thus, the Iron Man, as the operatives fitly call it, sprung out of the hands of our modern Prometheus at the bidding of Minerva—a creation destined to restore order among the industrious classes. . . . This invention confirms the great doctrine already propounded, that when capital enlists science in her service, the refractory hand of labor will always be taught docility.[1]

From the beginning of mechanization, with the invention of the Jacquard automatic loom, the self-acting mule, and other semiautomatic equipment, this management theme has echoed in the minds of inventors, including the earliest pioneers of computer automation. Thus Charles Babbage, the father of the modern computer, emphasized in his book on the economy of machinery and manufacturing (1832) that a "great advantage which we may derive from machinery is from the check which it affords against the inattention, the idleness, or the dishonesty of human agents."[2]

In the late 1940s, control engineers at MIT (who had just completed a rolling-mill control system designed to enable Bethlehem Steel management to eliminate "pacing" by workers) turned their "fertile genius" to the metalworking industry. The ultimate result of their efforts, "numerical control" (NC), reflected management's twofold objective and set the pattern for all subsequent development of what are now known as computer-aided-manufacturing systems. As its very name suggests, control was and remains its essence, not just management control of machines but, through them, of machinists as well.

With numerical control, there was a shift of control to management. The control over the machine was placed in the hands of management.

I remember the fears that haunted industrial management in the 1950s. There was the fear of losing management control over a corporate operation that was becoming ever more complex and unmanageable. Numerical control is restoring control of shop operations to management.

Numerical control is not a strictly metalworking technique; it is a philosophy of control.

Numerical control has been defined in many ways. But perhaps the most significant definition is that [it] is a means for bringing decision-making in many manufacturing operations closer to management. Since decision-making at the machine tool has been removed from the operator and is now in the form of pulses on the control media, [NC] gives maximum control of the machine to management.

There was little doubt in all cases that management fully intended to transfer as much planning and control from the shopfloor to the staff office as possible.

The fundamental advantage of numerical control has been spelled out: it brings production control to the Engineering Department.

In recent years, manufacturing industries in the U.S. have accelerated the move toward automating their operations. Factors that have motivated this move include the need to increase productivity, the high cost of labor, competition from abroad, and the desire for closer management control over production operations.[3]

The pattern is clear enough. At the dawn of the industrial revolution, Andrew Ure had boasted of how "in the resources of science, capitalists sought deliverance from [their] intolerable bondage" to the wit and will of the work force.[4] And the same is true at the dawn of this second, computer-based, industrial revolution. Making explicit the management dream of progress without people, an engineer from the Arthur D. Little consulting company wrote excitedly to MIT after viewing an early demonstration of numerical control that the new technology signaled at long last "our emancipation from human workers."[5]

Now, engineers have objectives of their own that neatly complement and innocently approximate those of management: they want to create an error-free system that will operate with a high degree of certainty in a manner perfectly faithful to the intentions of the designer. With this end in view, a "closed system," engineers need not deliberately seek, or even be aware of, the management goals they so consistently serve.

Above all, engineers want to eliminate not particular human beings but the more abstract possibility of "human error." So, they design systems that preclude as much as possible any human intervention. This is called "idiot proofing." This engineering equivalent to management's worker-proofing betrays a cynical view of human beings (not to mention an elitist and derisive view of subordinates). Any chance for human intervention (by workers) is negatively assumed to be a chance for error rather than, more positively, a chance for creativity, judgment, enhancement. Like other engineering habits, it reflects the engineer's privileged position in the industrial power structure. It is his relative power, not his scientific training, that enables and encourages the engineer to design systems to be operated by "idiots."

If engineers designed a machine that they would be operating, they would certainly leave ample room for their own later involvement in the process. Thus, if there were a law requiring all machine designers to operate their own equipment for five years after it is installed on the factory floor, there would be a revolution in engineering design.

Now, as it turns out, in the historical evolution of automated machine tools there actually have been attempts to do something like this but, despite their technical and economic promise, they all ultimately fell victim to the managerial obsession with control and its engineering counterpart, the quest for an error-free design. A series of such efforts, described in great detail in the chapter "The Road Not Taken" of my book *Forces of Production,* showed that their creators shared a more respectful appreciation of the talents, knowledge, and resourcefulness of shopfloor workers and an understanding of their vital role in efficient, quality production. Thus men such as Eric Leaver, one of several inventors of record playback control; F. P. Carruthers, designer of the Specialmatic control; and David Gossard, creator of the Analog Part Programming system—to name just a few—endeavored to build machines for machinists rather than for idiots. The aim was to take advantage of the existing expertise, not to reduce it through de-skilling; to increase the reach and range of machinists, not to discipline them by transferring all decisions to management; to enlarge jobs, not to eliminate them in pursuit of the automatic factory. Predictably, perhaps, but at any rate consistently, such alternative approaches remained stillborn.

Thus, NC became the dominant and, ultimately, the only technology for automating metalworking—the sole survivor and hence seemingly the best history had to offer, in the Darwinian view of technological progress. But this result was not the outcome of some natural selection by technical reason. Rather, it was the product of political selection by those powerful few seeking to retain and enlarge their social control, in league with those technicians who seek perfection in a world of idiots. In a machine shop in Lincoln, Nebraska, the ultimate fulfillment of these interwoven impulses, short of the totally automated workerless factory, was finally achieved. There, the NC equipment is run by a mentally handicapped operator with a maximum intelligence of a twelve-year-old child. According to the *American Machinist,* this man was selected for the job "because his limitations afford him the level of patience and persistence to carefully watch his machine and the work that it produces." "His big plus," the shop's manager enthusiastically explained, "is that he will watch the machine go through each operation step by step. . . . He unloads every table exactly the way he has been taught, watches the [NC machine] operate, and then unloads. It's the kind of tedious work that some non-handicapped individuals might have difficulty coping with."[6]

THE MILITARY MENTALITY

A British machinist quipped recently that management is just a bad habit inherited from the military and the Church, and there is much truth in the observation.[7] Whatever has been the role of the Church, the military mentality has always played a central role in the technological development of U.S. industry, from mining and metallurgy to shipping and navigation, from interchangeable parts manufacture to scientific management.[8] As the Army and Navy have been the major movers in the past, the Air Force has led the way in our time (the Marines, apparently, have been otherwise occupied).

Electronics, computers, aerospace systems, cybernetics (automatic control), lasers: all are essentially military creations. And when some of these war-generated technologies were brought together to automate the metal-working industry, the military was once again the driving force. From the late 1940s to the present day, the Air Force has been and remains the major sponsor of industrial automation. With regard to numerical control, the Air Force underwrote the first several decades of research and development of both hardware and software; determined what the technology would ultimately look like by setting design specifications and criteria to meet military objectives; created an artificial market for the automated equipment by making itself the main customer and thereby generating demand; subsidized both machine tool builders and industrial (primarily aerospace) users in the construction, purchase, and installation of the new equipment; and even paid them to learn how to run it. Numerical control was just the beginning of Air Force involvement in the automation drive. The Air Force NC project had global significance. On a recent visit to a locomotive factory in Prague, I was surprised to find the Air Force NC programming system in use even there. Before long, this single project had evolved into the more expansive Integrated Computer Aided Manufacturing Program. More recently, ICAM became the still more ambitious and diversified MANTECH (manufacturing technologies) programs, designed to promote the computer-automated approach to manufacturing not only in industry but in universities as well. "The Air Force automation programs were established to force development of the technology," an ICAM program director explained several years ago. "Factories of the Future [the Air Force's latest boondoggle] are being designed to serve as models for U.S. industry in which computers and machines can be made to work together with little human intervention."[9]

The effects of this military involvement reflect the peculiar characteristics of the military world. First and most obvious is the military emphasis upon command, the quintessence of the authoritarian approach to organization. The intent is to eliminate wherever possible any human intervention between the command (by the superior) and the execution (by the subordinate). An array of men behaving like machines is readily replaced by an army of machines in the military outlook. This command orientation neatly complements and powerfully reinforces the managerial obsession with control.

The second characteristic of the military mentality is the focus upon performance above all else, reflecting the mission-oriented priorities of "combat readiness" and "national security." This fixation on performance renders all else secondary, and fosters an industrial outlook that is more or less cost-indifferent. This explains the tendency toward waste, extravagance, and excess that marks so much military-sponsored effort and it explains also why at so many U.S factories today, while the American flag still flies overhead, Japanese machines are in use within. Preoccupied with meeting the exaggerated performance specifications of their primary customer, the military, U.S. equipment manufacturers have essentially priced and designed themselves out of the domestic commercial market. (In 1978, the United States became a net importer of machine tools for the first time since the nineteenth century).[10]

Finally, the military's extravagance, however damaging to competitive industry, has proved extremely attractive to technical people, who are drawn to this "anything goes" atmosphere where they can try out their latest dreams. Fully one-third of the nation's technical manpower go to work for the military because it offers them the biggest technological playground. The money is the major incentive, of course, but not the only one.

Equally important are the technical enticements. With its nearly unlimited resources, the military offers technical people the often unique opportunity to work with state-of-the art technologies at the cutting edge of development and the chance to dream expansively in pursuit of elegance and sophistication without regard to cost and other mundane practicalities. Moreover, with its rigidly defined chain of command and closely regulated environment, which guarantee a high degree of certainty and predictability, the military offers technical people almost laboratory conditions in which to try out their authoritarian designs. The military, in short, is able to indulge the collective enthusiasms and compulsions of technical people, which can be at once exhilarating and dangerous.

ENTHUSIASMS AND COMPULSIONS

As everyone has experienced at one time or another, technical challenges can be highly seductive. It is not unusual to get emotionally involved when trying to make something work, whether the challenge is manual or intellectual. You skip dinner, ignore the calls of nature and other people, push on into the wee hours, driven, possessed, determined. There is a delight in it, a passion—and a blindness. You can hardly tolerate interruption or delay, much less interference, and you get so you would almost kill to get the damn thing to work.

Of course, such emotional enthusiasm is the wellspring of creativity and can often be inspiring and enriching. But when it is indulged beyond reason, in defiance not only of personal health but of the larger social welfare as well, it becomes madness. Let us take a closer look at some of these enthusiasms and compulsions, the ones that underly automation madness.

To begin with, there is the shared ideal of a world without people, an image that affects the way these people view the activities around them; that is, their imagination distorts their perception. Consider, for example, the perceptions of Andrew Ure, the nineteenth-century authority on manufacturing who was cited earlier. When Ure examined early textile factories, this is what he saw:

I conceive that this title—Factory—in its strictest sense, involves the idea of a vast automaton composed of various mechanical and intellectual organs, acting in uninterrupted concert for the production of a common object, all of them being subordinated to a self-regulating moving force.[11]

Reading this passage from Ure's *Philosophy of Manufactures,* it is easy to forget that these early factories were teeming with people, people who magically disappear in Ure's description. This is a common (mis)perception of technical enthusiasts. Nearly a century and a half later, Dr. C. C. Hurd, Director of Applied Science at IBM, offered a strikingly similar observation:

It seems to me that the most useful analogy which I can see for the assembly line is that the assembly line—or, more generally, a complete production line—is like a computing machine.[12]

Whereas Ure entered a factory full of workers and saw only a vast automaton, a self-acting machine, so Hurd (in the manner of his like-minded ancestor Charles Babbage) saw only a computer. The power of abstraction of such men blinded them to the actual human realities of production. And when these realities rudely but invariably interfere in their reverie, they are viewed with contempt and arrogant impatience. For technical people absorbed by such imaginings, the ideal often becomes more real than reality itself: a fantasy of a perfectly ordered universe to which the world of people must be forever adjusted. The attractions of this idealized world of machines and computers are clear enough; it is above all a clean world, controlled, predictable. But there is more to it: if such a vision fulfills a deep-seated desire for order, it also satisfies an enchantment with things that are at once animated and artificial, almost lifelike in their autonomy, on the one hand, and yet under nearly complete (albeit remote) control, on the other. For these dreamers, there is sheer delight in such a spectacle. Where does this enthusiasm, this intrinsic fascination with automation come from?

One likely clue is the fact that this fascination with automation, an ancient obsession, has always been a peculiarly male preoccupation and remains so today. There is no evidence suggesting that women have shared this keen interest in what might be termed the artificial creation of life. A decade ago the British government commissioned a study of artificial intelligence (a technical field central to automata design). The final report of this study, authored by the eminent Cambridge University mathematician Sir James Lighthill, contains this provocative passage.

It has sometimes been argued that part of the stimulus to laborious male activity in creative fields of work, including pure science, is the urge to compensate for lack of the female capability of giving birth to children. If this were true, then building robots might indeed be seen as the ideal compensation. The view to which this author has tentatively . . . come is that a relationship which may be called pseudomaternal comes into play between a robot and its builder.[13]

Of course, all of this is just speculation, yet warrants serious attention. For whatever it is that is driving men to automate, it is also driving us all in the same direction, a direction in which we might not really want to be heading, or at least not at the current velocity. What is the goal? What is the hurry? There is something out of control here, something almost transcendent (not to mention socially irresponsible). "The automatic factory is like the Holy Grail—something you approach but never reach," Phillip Villers, The President of Automatix, Inc., solemnly observed in 1983.[14] *Fortune* magazine put it even better thirty years earlier, just as the automation advance began to accelerate:

> In the nature of things, man will create an automatic factory as he climbs Mt. Everest and aspires to reach the moon, for reasons no one has ever clearly expressed. Except that he is a man.[15]

Behind the Economic Screen

Our ideology of automatic advance tells us that the excesses of technical enthusiasts are corrected for by the economic screen of automatic technological progress—by the no-nonsense economic rationality of profit-seeking businessmen and by the ever-dependable self-regulating mechanism of the market. Would that it were true, for here too reality belies the Darwinian assumptions of our mythology.

THE ALL-TOO-HUMAN BUSINESSMAN

Our ideological image of the businessman is a caricature of a hard-working, practical-minded "economic man," guided by sober cost accounting and the pressures of competition and supply and demand, and intent above all upon making a buck. This stark abstract image, too inhuman to be real, is purveyed by the press, by public relations agents, and by businessmen themselves (not to mention by economists of all political persuasions) because it conveys the impression that business itself is abstract and therefore beyond our control, an objective, inevitable force in our lives rather than merely a mad scramble of greedy and familiar people. And this objective appearance only reinforces our idea of that automatic destiny we call progress. Businessmen, we are reminded almost daily, have a job to do—they have no time for dreamers. Businessmen are straightforward, down-to-earth, simple to understand. They are predictable, so we can count on them. They are ruthless in their blind pursuit of profit, so we can trust them. This is why we have such confidence in the economic screen—this sublime self-interest takes on a life of its own. Just as we have believed that technical people are merely agents of an abstract autonomous technical rationality, so do we here assume that businessmen are simply agents of an abstract and autonomous economic rationality. The engineer evaluates the machine by asking, Will it make widgits? The businessman looks again and asks, Will it make money? Given their dedication, we cannot fail to get a technology that is both technically and economically viable.

But, alas, as we have seen in the case of the technical people, appearances are deceiving. The businessman, too, is more human than our caricature allows. He too has dreams and delusions, enchantments and enthusiasms, flights of fantasy: "U.S. companies are on the verge of achieving a dream . . . [of] manufacturing enterprises where push button factories and executive suites, no matter how physically remote become part of the same computerized factory."[16] Of course, businessmen want to make a profit, believe that their actions will have that result, and justify those actions always in economic terms. But this is far from the whole story. Let us take a closer look at what moves the businessman and, through him, the advance of automation technology.

For one thing, justifications are not the same as motivations. In reality, the former tend merely to obscure the latter and serve instead as rationalizations for actions taken for unstated reasons. Despite the authoritative appearance of such economic calculation—with every cost estimated down to the last decimal point—there is typically less there than meets the eye.

According to neoclassical economics, businessmen decide whether to invest in machinery by comparing its cost with that of labor. If the cost of machinery is less than the cost of labor, they will invest in machinery and if the cost of machinery is more than the cost of labor, they will stick with labor. A young Harvard economist undertook to test this theory in the field by conducting a survey of some sixty factories in the New England area. In each case he identified and talked with the people who actually made such purchasing decisions, and tried to find out how they did it. He discovered, first of all, that the majority of these people had technical backgrounds; they were engineers. He also found out that their actual purchasing behavior differed from what the theory suggested. When the cost of machinery was lower than that of labor, they bought machines, but when the cost of machines was higher than that of labor, they still bought machines (and sometimes fudged the justification accordingly). The economist concluded that there was a bias in favor of machinery (or against labor) on the part of these technically trained functionaries. Their enthusiasm for machinery was the major determining factor, not careful relative-factor analysis.

During factory visits related to my own study on automation, I had gotten used to the unexpected. I remember talking with a man who installed a particular type of numerical control equipment. He told me a strange story of how in one shop he noticed a row of assorted castings lined up against the wall next to where he was installing the new machine. After asking around, he learned that these machine jobs had been used to justify the purchase of his machine. He was astonished, because his machine could not make machine castings; it was a turret punch press that worked only on sheet metal.

In other shops, where the jobs at least matched the capability of the equipment, I learned that, although managers consistently boasted about how the new equipment increased productivity, they never seemed to have any hard evidence. I found out also that productivity is itself a slippery concept, hard to define much less measure, and also that so-called "creative accounting" among divisions of a company spread the data around in such a way as to make it almost impossible even to make an educated guess. More amazing still, I discovered that there were almost never any audits done on equipment, to assess after the fact whether expected benefits (used to justify purchase) had actually been realized; apparently few people want to learn from or even to document their errors, so, once a machine is put in concrete, it usually stays there—and so does the next one. Although the vaunted economic rationality of businessmen sometimes comes into play, more often than we suspect other impulses—cultural, political, psychological—underlie decisions on technology. Only after the choice has been made, only after the equipment has been installed, is there any serious effort to render it economically viable—with mixed results, as we will see, and usually at public expense.

As the Harvard economist's study indicated, the promoters of new technology within companies are typically people with technical backgrounds and the enthusiasm for technology that goes with it. In 1981 Donald Garwin of Arthur D. Little, Inc., conducted a study of automated flexible manufacturing systems (FMS) and found that "management is usually sold on the idea to use such systems by an engineer in the company who is enthusiastic about the technology. Cost justifications play a secondary role. The more sophisticated and fascinating a machine is, the less management is likely to quarrel over dollars."[17]

Joseph Engelberger, founding president of Unimation, Inc., the first major U.S. robot manufacturer, knows this well. As a super salesman of industrial equipment, he well understands what sells machines and is candid about the enthusiasms of his customers and their preoccupation with such intangibles as social status. Recently he remarked to the press

> I don't think a guy will be able to go into his country club if he doesn't have a CAD/CAM [computer-aided manufacturing and design system] system in his factory. He's got to be able to talk about his CAD/CAM system as he tees off on the third tee—or he will be embarrassed.[18]

In addition to these concerns of individual managers, there is a collective phenomenon at work as well, a sort of herd instinct. Like a run on a bank or the stock market, someone starts it and before long everyone is doing it out of desperation. Managers feel they must automate because "everyone's doing it," out of fear that they will be undone by more up-to-date competitors (a paranoia encouraged by equipment vendors). There is this vague belief that the drive to automate is inevitable, unavoidable, and this belief becomes a self-fulfilling prophesy. In the stampede, meanwhile, there is very little sober analysis of costs and benefits. "Everybody and his brother believes that FMS is the only way to fly," the trade magazine *Iron Age* reported. "Yet, there isn't a single FMS in the U.S. that operates the way it was intended to."[19] Companies are buying equipment helter skelter without thinking about how they want to use it," one Boston University business school professor told the *Wall Street Journal* (11 April 1983).[20] In short, this trance that we are all in—the feeling that there is some inevitable force called technological progress and you have to hop on or get run over—executives are in too.

THE MARKET MIRAGE

If you can't trust the technical people and now you can't trust the businessman, who or what can you trust to keep technological progress on course? Happily, there's still the market, that mysterious yet infallible mechanism that magically makes everything work out in the end. Just as it miraculously transforms the individual pursuit of self-interest into the larger social good, so it consistently corrects for the excesses and errors of individual businessmen by forcing them into bankruptcy and out of the picture. Only the sober, smart, and savvy survive and thus, finally, in this competitive court of last resort, our Darwinian assumptions of natural selection are upheld. Not quite.

The convenient fiction of the market was a nineteenth-century propaganda invention created by upwardly mobile bourgeoisie to challenge the economic power of the state and thereby extend the range of their exploitation.[21] In reality, the "free" market has never truly existed because businessmen have always used all the political power at their disposal to influence events in their interest: they used the state to create the "free market" in the first place by doing away with regulations protecting workers and consumers; they enacted various protective devices for themselves, from state-chartered and subsidized corporations and tax incentives to military support of enterprise and, of course, tariffs. And the same is true today, where the role of government in the economy is greater than ever before. The supposedly self-regulating mechanism of the competitive market is easily overwhelmed by the power of the state as both underwriter of enterprise and largest customer. In the case of automation, as we have seen, the state, especially the military, has played a central role. Not only has it subsidized extravagant developments that the market could not or refused to bear but it absorbed excessive costs and thereby kept afloat those competitors who would otherwise have sunk. As one Air Force official candidly observed.

We have contractors with divisions set up just to get Air Force projects. We're keeping them alive. People are automating for automation's sake in several cases. There is no good reason, there is no good justification— and in fact it may be detrimental. We work with companies whose job it is to implement these advanced technologies, and if they can get a project from the Air Force, regardless of its real payback, they keep in business.[22]

It is thus no accident, for example, that the machine tool builders trade association moved its headquarters from the Midwest center of the industry to Washington, D.C., home of its major customer, the Department of Defense. Nor is it an accident that the defense-related industries are the ones with the most automation. These industries, moreover, are expanding along with the military automation programs, as more and more businesses rush to this state-supported sanctuary to escape the unpredictable vicissitudes of the market. At the same time, the military automation programs are today being matched by those of civilian agencies such as the Department of Commerce and the National Science Foundation. All have now become the publicly funded pushers of automation madness, charting a course and prompting a pace that no self-adjusting market, had it existed, would ever have tolerated.

The economic power of gigantic multinational corporations, some of which exceed the scale of governments, allows managers to carry costs, and conceal costs, that would cripple other firms. And their sheer economic (and thus political) muscle enables them to corner markets, intimidate or "acquire" competitors, and thereby distort beyond measure the real costs of doing business. The relationship between corporate profit and economic production is becoming more incidental every day. The corporate automation drive is just one case in point. Not surprisingly, it is the giant firms that are the leaders in this drive and it is difficult if not impossible to evaluate their returns.

Thus, the market panacea turns out to be just one more mirage that evaporates upon closer inspection. No automatic guarantor of economically sound technological progress, it is instead yet another ideological camouflage for political power. Perhaps it is time now to leave Darwinism to biology where it belongs and to start looking at this important matter of technological progress more critically, because it has serious consequences for us all. Having overcome the first half of the mythology of automatic progress by examining the human and social drives behind technological development itself, let us turn now to the other half, to examine the consequences of these drives—where they have led and are still leading us. According to our inherited ideology of progress, automatic technological progress leads automatically to social progress. But if, as we have seen, there is really nothing automatic about technological progress itself, what can we expect in the way of social progress? Could this turn out to be a story without a happy ending?

A SECOND LOOK AT SOCIAL PROGRESS

In 1952, at the dawn of the so-called age of automation, a Pratt and Whitney engineer waxed eloquent about its promise. A pioneer of automation himself, he used the central metaphor of cybernetics to express his quasi-religious faith in the automatic beneficence of technological progress.

> I don't think we are consciously trying to ease the burden of our workers, nor consciously to improve their standard of living. These things take care of themselves. They have a feedback of their own that closes the loop automatically.[23]

This faith takes many forms in our culture and is rarely ever articulated with any clarity, much less with logical rigor. But in outline, this ill-defined faith could be called the beneficent circle of prosperity. It goes like this—people with money are offered incentives (the chance to make more money) to lure them to invest in new, improved plant and equipment (so-called innovation). This innovation automatically yields increased productivity and, hence, lower costs and prices, which results in greater competitiveness. Finally,this enhanced competitiveness necessarily brings about what Adam Smith, the great eighteenth-century philosopher of capitalism, called the "wealth of nations": economic growth, jobs, cheap and plentiful commodities—in short, prosperity. Let us now take a closer look at this magic circle, to examine these assumptions on their own terms. For when we do, we will see that the causal chain is, in reality, ambiguous at each link, and the end result is not exactly what we blithely assume it must be.

What Goes Around Comes Around

To begin with, the assumption that rich people will invest in new means of production if given sufficiently lucrative incentives presupposes that they would not do so voluntarily without such inducements. As such, it is itself a tacit recognition of the inadequacy of the market as a stimulus to development. Furthermore, it rests upon the prior assumption that people simply follow their pocketbooks, which, while largely true, is not, as we have seen, the whole truth. More to the point here, to the extent that people do strive primarily for the highest return on their investment, there is no guarantee that they will invest in new means of production if other, more profitable routes are available, even given incentives. It certainly appears to be the case in our own day, judging by patterns of investment, that, however vital production remains to any economy, it has become relatively less attractive as a way to make money. In the past, production was the point of entry for ambitious capitalists closed to more established avenues of power and wealth. Today, this historical connection between capitalism and production appears to be fading as an increasing proportion of investment is diverted into nonproductive areas of the economy, such as real estate and financial speculation. Indeed, there seems to be a scramble among capitalists to get out of the messy, troublesome business of actually producing something for society whenever they have an opportunity, a trend toward disinvestment that is leaving a trail of debris—closed plants, idle workers, ghost towns—in its wake. In an age when oil companies invest in circuses, manufacturers invest in real estate,and the steel industry deliberately and openly abandons the production of steel, there is good reason to reconsider this first link between investment and innovation.[24]

The second causal link—between innovation and productivity—is more difficult to assess but, judging from the case of automation, it too appears to be ambiguous at best, with no evidence solid enough to rely on. When investment does in fact generate innovation, does such innovation necessarily yield greater productivity? The assumption here is that the return of profits to the investor will be matched by more and cheaper goods for society. This assumption, of course, is the cornerstone of apologies for capitalism, its central tenet of legitimation. But today even the business press has begun to back away from this claim. After conducting a poll of industry executives on trends in automation, *Business Week* concluded in 1982 that "there is a heavy backing for capital investment in a variety of labor-saving technologies that are designed to fatten profits without necessarily adding to productive output."[25]

But few are able to confront, much less draw the correct implications from, such ideologically disorienting disclaimers. Thus, efforts to document productivity gains continue to abound, despite the fact that the concept of productivity is hard to define and the reality equally difficult to measure. For one thing, there is very little hard information available.

Even my more optimistic colleagues at MIT concluded from their preliminary investigation of industrial automation that "after-the-fact analysis of the actual economic impact of a process of automation is rarely carried out; the result is a loss of data necessary to inform subsequent decisions about automation."[26] After abandoning my own attempts to get hard data I resorted to the less "scientific" approach of asking people in the factories I visited. Yet this also proved problematic. According to the managers, every new machine increased productivity, often dramatically. After a while the plausibility of this too consistent claim began to wear thin, especially given the fact that the shopfloor people who worked with the equipment invariably told a different tale, of downtime and disasters. The reality of productivity, however one defined it, remained ambiguous; in recent years, even the business press has begun to acknowledge such "mixed results." "The results are mixed," the *Wall Street Journal* reported in 1980.

As technology soars, users struggle with the transition and unsuitable machines; computerized equipment often doesn't work the way it's supposed to, the new equipment is more fragile than the old-fashioned equipment . . . and problems with the software used to run the equipment are even more prevalent.[27]

[Companies] get the biggest, fastest, sexiest robot, when the plain truth is that in most cases a very simple piece of equipment could do the job. . . . [Companies] don't so much make mistakes as learn that it's going to take two or three times as much money and time as they thought to get the system working.[28]

Economists typically respond to such news of technical unreliability and economic uncertainty with calm confidence, arguing that industry is now just going through the learning stage and that, after a while, the expected productivity gains will be realized. But this wishful thinking—which disarms critics and forever defers judgment—is perhaps too sanguine and smug, as the experience of the banking industry, which has had more time to move along the "learning curve," now suggests. A recent study of the economics of innovation in this area, reported by *Computing Canada* on 16 May 1985, found that "senior executives of banks are generally disappointed in the return on their investments in technology."

The inability to use technology to achieve lasting competitive advantages, and the failure to achieve expected economic returns through reduced operating costs, were among reasons given by senior management of 200 major banks and financial institutions in 26 countries surveyed by the management consulting firm.[29]

At least some inside observers have begun to acknowledge a similar disparity between religion and reality in the industrial automation experience. Henry Miley, retired Air Force general and Director of the American Defense Preparedness Association, expressed his concerns about the economic returns on military investment in automation, in testimony before the House of Representatives in 1980.

What concerns me is that when I get up and raise my Yankee voice and say, can I go out to some factory and put my hands on an item that is being produced more cheaply now than it was five years ago because of the [Air Force automation programs], I get kind of a confused answer. When I ask the bottom line question, is the [product] now cheaper than it was two years ago because [of the Air Force program], the answer was, well, no.[30]

Apparently, the ambiguity of results here reflects the overriding performance imperatives of the military, not to mention the enthusiasms of technical people and the preoccupation of management with control and the fast hustle. But because these factors pervade those very industries that are now undergoing automation, there is certainly reason to generalize from General Miley's assessment. Has anyone noticed any decline in prices lately?

Given these mixed results, one might expect any rational person to abandon the ready assumption that all innovation increases productivity and to seek more sober assessment. Yet any such effort to do a careful and critical assessment, to try to separate the wheat from the chaff, immediately invites the charge of "Luddism," enemy of progress. Ironically, in our day, the demand for greater productivity has become a revolutionary slogan rather than a paean to capitalism, because it exposes the soft, ambiguous underside of our seemingly authoritative economic justifications for domination.

For example, most analysts and industry accountants measure productivity as output per person hour, that is, product per unit of labor time, where labor means hourly production (direct) labor and time is hours on the job. (The engineers I have talked to typically use a standard number to estimate the cost of labor time although none of those interviewed knew how it was derived.) An overriding assumption of almost all discussion about automation is that productivity increases result from the substitution of machines for hourly production workers. That is, a reduction in factory jobs is ipso facto understood to mean a gain in productivity. Moreover, the effort to reduce the work force by managers is universally understood to reflect an interest in increasing productivity. A closer look coupled with a genuine concern for productivity would challenge this assumption and perhaps reveal other motivations at work. For, as Thomas Gunn of Arthur O. Little, Inc., argued in 1982,

"Direct labor accounts for only ten to twenty-five percent of the total cost of manufacturing. . . . It is not clear that even a total replacement of blue collar workers by robots would [by itself] have much effect on the output of the factory or the cost of its products.[31]

John Simpson, Director of Manufacturing Engineering at the National Bureau of Standards, took this same message a bit further.

In metalworking manufacture, direct labor amounts to roughly 10 percent of total cost, as compared to materials at 55 percent and overhead another 35 percent. Yet, as of 1982, management was expending roughly 75 percent of managerial and engineering effort on labor costs reduction, as compared to 15 percent on materials cost reduction and 10 percent on overhead cost reduction. This is a striking disparity.[32]

As *Business Week* discovered in its 1982 survey of executives, few managers anticipated much use of the new equipment to displace management, even though such reduction in overhead, as Simpson suggests, would no doubt serve the goal of increased productivity.[33] But is it really the goal? The automatic causal link between innovation and productivity is ambiguous, and not only in terms of actual results but perhaps also in terms of intentions. Whenever managers are able to use automation to "fatten profits" and enhance their authority (by eliminating jobs and extorting concessions and obedience from the workers who remain) without at the same time increasing the social product, they appear more than ready to do so.

The next weak link in the causal chain connects productivity with competitiveness. The assumption is that greater productivity results in lower operating costs and thus lower prices and that this increases demand and enhances competitive strength. As we have seen, the increases in productivity due to automation are hard to assess with confidence. And it appears also that when there have been productivity gains and, presumably, lower operating costs, these have rarely resulted in any reduction in prices. Wherever the gains are going, it is not to the consumer. But, for arguments sake, let us suppose that there have been lower costs and also lower prices. Does this guarantee greater competitiveness? Not really.

The truth of the matter is that competitiveness is tied less to operating costs and prices than it is to product quality, product design, marketing strategies, service—as successful Japanese firms especially have demonstrated again and again. The ability to produce something more cheaply without these other factors might result only in the swelling of inventory, not sales. The simple assumption that lower costs through increased productivity results in more sales is naive and misleading. The Canadian study of banking automation, it should be emphasized, reported that managers acknowledged their "inability to use technology to achieve lasting competitive advantages." Instead, the investigators found that banks were introducing new technologies merely to "appear to be competitive."[34] Here as elsewhere appearances can be deceiving.

Finally, to close the circle, there is the last assumption that increased competitiveness of firms results in prosperity for all. At the time Adam Smith formulated his ideas about how to increase the wealth of nations, in the late eighteenth century, he had good reason to assume some parallel between the prosperity of a company and the prosperity of its homeland, and the correlation continued to hold for several centuries thereafter.[35] But it no longer does. For today, most major manufacturing firms are multinational not only in terms of their market but also in terms of the scale of their production operations. This global scale of operations is matched by a relatively unimpeded mobility. Firms have the ability to transfer production from one country to another, to close a plant in one and reopen it elsewhere, to direct and redirect investment wherever the "climate" is most favorable. This mobility has resulted in a rupture between the health of a corporation and the prosperity of any one "host" nation (including its home base). That is, even when innovation and productivity do actually combine to increase the competitiveness of corporations, such competitiveness is no panacea, no guarantee of prosperity. Indeed, it has only better enabled the corporation to play one work force off against another in pursuit of the cheapest and most compliant labor (which gives the misleading appearance of greater efficiency). Moreover, it has compelled regions and nations to compete with one another to try to attract investment by offering tax incentives, labor discipline, relaxed environmental and other regulations, and publicly subsidized infrastructure such as roads and sewers. Thus has emerged the great paradox of our age, according to which those nations prosper most (attract corporate invest-ment) by most readily lowering their standard of living (wages, benefits, quality of life, political freedom). The net result of this system of extortion is a universal lowering of conditions and expectations in the name of competi-tiveness and prosperity.

Thus, our circle of prosperity, upon closer inspection, appears less than compelling. And ideological imaginings alone cannot contradict the ambigu-ity of this promise. "The results," as the *Wall Street Journal* deftly under-stated it, "are mixed." But if the economic returns on automation remain ambiguous, the social consequences of automation do not. Unfortunately, they are all too clear.

Progress for Whom?

In their study of industrial automation, my MIT colleagues found that, just as there was little reliable data or certainty about the technical and economic viability of automation, so too there was "no commonly accepted calculus for assessing the societal costs and benefits of automation." They concluded, therefore, that "the absence of such a calculus and of data to support it seriously weakens arguments in favor of automation."[36] But, unfortunately, neither data nor arguments have ever had much of a role in the drive to automate. Rarely challenged for either, the evangelists of automation have never had to marshal evidence or formulate arguments to defend their posi-tion; the power of their hegemonic ideology of progress has alone been sufficient to carry their campaign. The burden of proof, rather, has been borne alone by the critics of this campaign. Only the critics have had to come up with evidence and argument, usually to have it dismissed without a hearing because it did not conform to the commonly held ideological presup-positions. Thus, it has been difficult for critics to provoke any serious—and much needed—debate.

The typical approach has been to speculate about the future, to estimate the number of jobs that will be lost or created. But such a crystal-ball approach is little more than a guessing game, and a biased one at that because the future looks grim, or rosy, depending upon who's looking and who's paying for the forecast. Every critical forecast is matched by an optimistic one. A more meaningful approach for estimating where we are headed is to examine historically where we have already been. Automation is not new. The term itself was coined in 1947 to refer to automatic transfer machinery in the automobile industry, and the introduction of computer-automated equip-ment has been going on for some thirty years. The returns are in on this experience. What do they tell us, and what are their implications? In addition to such a historical approach to this question, it is helpful also to be more specific about exactly whom it is we are speculating about. The impact of automation on society boils down, ultimately, to the impact of automation on particular people. There is no common calculus for assessing the societal costs and benefits of automation precisely because these costs and benefits are not borne and enjoyed by the same people and one man's gain is another man's loss. Thus, in trying to assess the likely social consequences of this progress we must learn to ask: Progress for whom?

In contrast to an economic analysis, the results of a historical and class analysis of automation are hardly ambiguous: by and large, the gains have gone to those with power, at the expense of those without it. And, as this essay suggests, there has been nothing automatic or inevitable about this outcome. It has followed not from any natural selection process or technical or economic logic but, rather, from the political and cultural conditions that have prevailed (and continue to prevail). The differential impact of automation is readily seen in the history of U.S. manufacturing industry over the past thirty years, judging from the official (and probably optimistic) data of the Bureau of Labor Statistics and the Department of Commerce.[37]

First, during this period, the value of capital stock (machinery) relative to labor doubled, reflecting the trend toward mechanization and automation. As a consequence, although the rate of productivity gain declined by half, the absolute output per person-hour increased 115 percent. But during this same period, real earnings for hourly workers, adjusted for inflation and cost-of-living, rose only 84 percent. Thus, after three decades of automation-based progress, workers are now earning less relative to their output than before. That is, they are producing more for less—working more for their bosses and less for themselves.

Second, instead of seeing the number of working hours reduced in the wake of so-called labor-saving automation, workers have seen the average work week either remain constant at 40 hours or actually increase due to compulsory overtime and shiftwork imposed by a management intent upon getting the fullest utilization of their expensive new equipment (not to mention the additional hours resulting from the loss of paid holidays and sickdays extorted from unions by management with the threat of technological displacement). In short, labor-saving technologies have not been used to save worker's labor (meaning physical and mental effort) but rather to save capital labor (meaning workers, and wages). This double meaning of the word *labor* in a system in which one person can buy another person's labor power continues to confuse many observers.

Third, in the 1950s and 1960s, sociologists began to worry about the problem of "leisure"—what would workers do with the time made available by the shorter work week and less toil? But just as the shorter work week never materialized, neither did the leisure—except in the form of enforced, involuntary leisure known as unemployment. Instead of being relieved of effort, workers have been relieved of their livelihood. This human debris of progress was for a time obscured by the overall growth of a warspurred expanding economy, and some of it was temporarily absorbed by the government-subsidized enlargement of the so-called service sector. But, once this boom began to bust, the outlines of all this progress became more visible. The proportion of employment in manufacturing relative to the total non-agricultural employment had already declined by a third, and it appeared that this was just the beginning of the end.

In 1978, the international Organization of Economic Cooperation and Development (OECD) conducted a study of the impact of automation and concluded that "the evidence that we have is suggesting increasingly that the employment effects of automation, anticipated in the 1950s, are now beginning to arrive on a serious scale."[38] By 1982, as the economy continued to contract and the service sector itself came under an automation assault, even the traditionally sanguine General Accounting Office had to concede that "whether automation will increase unemployment in the long run is not known."[39] The Congressional Budget Office was less circumspect. In 1983, that office estimated that, by 1990, "a combination of automation and capacity cutbacks in basic industry will eliminate three million manufacturing jobs."[40] This time a new escape hatch was invented; not the service sector but the "high-tech" industries, especially the manufacturers of automation equipment, would expand to absorb those displaced. But before too long, even executives in those "high-tech" industries had begun to express their doubts. "If you look at the long-term unemployment forecast and couple it to the whole issue of retraining, the problem is bigger than . . . we think," Peter Scott, executive vice president of United Technologies Corporation, told the National Academy of Engineering in 1983.[41] David T. Kearns, chief executive officer of Xerox Corporation, also acknowledged that "at least in the short term, I think there's a very real possibility that technology will put more people out of work before it puts them to work."[42] Likewise, *Iron Age* predicted that "there will be massive displacements."[43] The number of new jobs created by high tech "will fall disappointingly short of those lost in manufacturing," *Business Week* finally concluded in 1983, citing Nobel Prize-winning economist Wassily Leontieff's more colorful, but no less austere, conclusion that automobile workers had about as much of a chance of getting jobs building robots as horses did building automobiles.[44] "All the talk about new technology creating more and more manufacturing jobs, with workers using higher and higher skills, is a figment of the imagination," labor economist Sar A. Levitan also concluded in 1984; "there is just not that much there."[45]

Thus far, then, the consequences of automation for workers are no cause for optimism. The loss of income relative to output, the constant forty-hour work week, and the rising specter of unemployment do not create a promising picture, as Leontieff (one of the few economists with the courage to tell it as it is) has explained.

> [The] value of capital stock employed per man-hour in manufacturing industries in the U.S. . . . has almost doubled since the end of World War II. . . . Since the end of World War II, however, the work week has remained almost constant. . . . Concurrently, the U.S. economy has seen a chronic increase in unemployment from one oscillation of the business cycle to the next. The 2 percent accepted as the irreducible unemployment rate by proponents of full-employment legislation in 1945 became the 4 percent of New Frontier economic managers in the 1960s. The country's unemployment problem today exceeds 9 percent [1982]. . . . Americans might have [absorbed] potential technological unemployment by voluntarily shortening the work week if real wages had risen over the past 40 years faster than they actually have. . . . Sooner or later, and quite probably sooner, the increasingly mechanized society must face another problem: the problem of income distribution."[46]

Again, progress for whom? As Leontieff suggests, the consequences have not been evenly distributed, have not been the same for everyone. For if the impact of automation on workers has not been ambiguous, neither has the impact on management and those it serves—labor's loss has been their gain. During the same thirty-year period of the age of automation, corporate after-taxes profits have increased 450 percent, more than five times the increase in real earnings for workers. To the extent that there have been tangible benefits from automation, they have gone in only one direction: up. This fact was made painfully clear recently by the telling behavior of the automobile industry. In 1983, as the industry recovered from its temporary slump, General Motors paid six thousand of its executives almost $200 million in bonuses, averaging more than what an average GM worker makes in a year. Ford, not to be outdone, paid its top forty-five executives a half million dollars each and its chairman $7.3 million (not a bad year). According to the *Los Angeles Times,* the record profits that made all this self-serving largesse possible resulted in part from the "introduction of modern equipment and sharp reductions in the automotive labor force."[47]

Once again, there was nothing automatic about these disparate outcomes. They followed from the social choices made by those who have had the power to choose. And, given this same constellation of forces, the future will more than likely be more of same: prosperity for the fortunate few and structural unemployment for the rest—precisely the starkly stratified have–have not society depicted by Kurt Vonnegut at the dawn of the automation age.

Calling Mr. Goodwrench

And still the snow-job-turned-sermon ideology of progress holds us under its stupefying spell, blinding us to this perilous prospect and conjuring up a prettier picture—automatically. But there are signs that at least some people have begun to see through this mystifying haze and to recognize more clearly what is at stake. Early in 1984, the Louis Harris opinion survey research organization published the results of an extensive public poll they had conducted on the impact of technology on society. They discovered that people viewed this thing called progress differently depending upon where they sat.

> The difference between the public and the corporate executives on the matter of robots is a startling 54 percentage points. The tension between social classes is unmistakable. By 39 points, corporate executives are more optimistic about factory automation than are the people who work in factories. In addition, executives are more optimistic than skilled and unskilled labor as a whole by 41 points. These figures represent a potentially combustible mixture.[48]

Apparently, then, people are beginning to see automation madness for what it is and to recognize the management sermon on progress as the snow job it has always been. "If they have the right to say yes to technology and then move, we have the right to say no and prevent them from moving; That's equality," Frank Emspak, a local union leader at a large GE plant in Lynn, Massachusetts, recently declared.[49] In other words, the progress of automation proceeds automatically at our expense only if, by our passivity, we allow it. Participation here demands defiance, defiance not only of the deceptive and disarming mythology of an automatic destiny, but also of the destructive designs of those who peddle it. Such defiance alone, of course, is not sufficient. But without it we will never regain the confidence or the power to take this very serious matter of progress back into our own hands, where it belongs.

NOTES

1. Andrew Ure, *Philosophy of Manufactures,* 3d ed. (New York: Burt Franklin, 1969), pp. 336–68.

2. Charles Babbage, *On the Economy of Machinery and Manufacturing,* 4th ed. (1835; reprint, Seamen, Ohio: Kelley Publications, 1963), p. 54.

3. Earl Troup, personal interview, 1977; Willard Rockwell, address to Western Metal and Tool Exposition, 11 March 1968 (North American Rockwell Corp., 1968); *American Machinist,* 25 October 1954; Nils O. Olesten, "Stepping Stones to NC," *Automation,* June 1961; Earl Lundgren, "Effects of NC on Organizational Structure," *Automation,* January 1964; *Iron Age,* 30 August 1976; *Industrial Engineering,* April 1984, p. 50.

4. Andrew Ure, *Philosophy of Manufactures.* (New York: Burt Franklin, 1969), p. 369.

5. Alan A. Smith to J. O. McDonough, 18 September 1952, NC Project Files, MIT Archives.

6. "Special Reasons to Hire the Handicapped," *American Machinist,* July 1979.

7. British machinist, quoted by Mike Cooley in talk at MIT, Winter 1979.

8. For a sustained examination of the role of the military in technological development, see Merritt Roe Smith, *Military Enterprise and Technological Change* (Cambridge, Mass.: MIT Press, 1985).

9. Dennis Wisnosky, quoted in Jerry Mayfield, "Factory of the Future Researched," *Aviation Week and Space Technology,* 5 March 1979, pp. 35–37.

10. For a penetrating examination of the negative effect of the military influence on commercial enterprise, see Seymour Melman, *Profits without Production* (New York: Alfred A. Knopf, 1983).

11. Ure, *Philosophy of Manufactures,* p. 1.

12. C. C. Hurd, quoted in "The Automatic Factory," *Fortune,* October 1953.

13. Sir James Lighthill, *Artificial Intelligence* (London: Science Research Council, April 1983).

14. Phillipe Villers (President, Automatix, Inc.), quoted in Dallas, "The Advent of the Automatic Factory."

15. "The Automatic Factory," *Fortune,* October 1953, p. 195.

16. *Business Week,* 13 December 1982.

17. Garwin cited in Vera Ketelboeter, "Where is Automation in Manufacturing Headed?" *Science for the People,* November/December 1981.

18. Engelberger, quoted in Gene Bylinsky, "A New Industrial Revolution Is on the Way," *Fortune,* 5 October 1981, p. 114.

19. *Iron Age,* cited in Tom Schlesinger, *Our Own Worst Enemy* (New Market, Tenn.: Highlander Center, 1983).

20. Stephen Rosenthal, quoted in "High Tech Track," *Wall Street Journal,* 11 April 1983.

21. For an enlightening history of the idea of the market, see Karl Polanyi, *The Great Transformation* (Boston: Beacon Press, 1944).

22. Gordon Mayer, quoted in Schlesinger, *Our Own Worst Enemy.*

23. J. J. Jaeger, quoted in "The Automatic Factory," *Fortune,* October 1953, p. 190.

24. For a thorough account of disinvestment, see Melman, *Profits without Production,* and also Barry Bluestone and Bennett Harrison, *The Deindustrialization of America* (New York: Basic Books, 1982).

25. Business Week/Harris Poll, *Business Week,* 13 December 1982.

26. "Preliminary Report on Industrial Automation," Center for Policy Alternatives, MIT, 1977.

27. "High Tech Track," *Wall Street Journal,* 13 April 1983.

28. Thomas Gunn, quoted in "High Tech Track," *Wall Street Journal,* 4 April 1983.

29. "Costs of Technology Outstrip Benefits for Banks," *Computing Canada,* 16 May 1985 (courtesy Rene Jacques, Local 240, UAW Canada).

30. Henry Miley, quoted in Schlesinger, *Our Own Worst Enemy.*

31. Thomas Gunn, *Scientific American,* September 1982, p. 129.

32. John A. Simpson, "The Factory of the Future," talk given at Rensselaer Polytechnic Institute, 3 June 1982.

33. Business Week/Harris Poll, *Business Week,* 13 December 1982.

34. "Costs of Technology Outstrip Benefits," *Computing Canada.*

35. Adam Smith, *The Wealth of Nations* (New York: E. P. Sutton, 1933).

36. "Industrial Automation," final report, Center for Policy Alternatives, MIT, 1978.

37. The statistics on labor productivity, wages, hours, and employment are from the U.S. Department of Labor, Bureau of Labor Statistics; the data on corporate profits are from the U.S. Department of Commerce, Bureau of Economic Analysis.

38. OECD forecast, cited in *The New York Times,* 5 July 1978, p. D1.

39. GAO forecast, cited in *The Washington Post,* 27 July 1982.

40. Congressional Budget Office forecast, cited in *The Wall Street Journal,* 3 April 1983.

41. Peter Scott (United Technologies executive), quoted in *NAS/NAE News Report* (National Academy of Engineering), May/June 1983, p. 23.

42. David Kearns (Xerox CEO), quoted in *Rochester Review* (University of Rochester), Winter 1984, p. 8.

43. "Factory of the Future," *Iron Age,* 25 February 1983.

44. *Business Week,* 28 March 1983.

45. Sar Levitan, quoted in *The New York Times,* 25 March 1984.

46. Wassily Leontieff, "The Distribution of Work and Income," *Scientific American,* September 1982, pp. 190 and 192.

47. *Los Angeles Times,* 2 May 1984.

48. *The Road After 1984* (New York: Louis Harris & Associates, Inc., 1983).

49. Frank Emspak, quoted in *Multinational Monitor,* March 1984, p. 18.

Engineering Birth:
Toward the Perfectibility of *Man?*

JOAN ROTHSCHILD

INTRODUCTION: PURPOSE AND ARGUMENT

In the heyday of the eugenics movement in the early part of this century, it was firmly believed that humans were at the top of the evolutionary hierarchy. It was also seriously proposed that these superior creatures could and should be "improved." Intellectuals and other opinion leaders joined with the scientific community to support this forward march of human progress. Natural selection would be helped along by selective breeding: encouraging the fit ("positive eugenics") and weeding out the unfit ("negative eugenics"). The biology of the eugenicists was soon shown to be faulty, their ideas discredited by a later generation as racist—and classist, sexist, and ethnocentric as well. But a key element of the belief system supporting the evolutionary theory and eugenics of the late nineteenth and early twentieth centuries did not die. What survives is its ideology of *human progress*. The ideology consists of: the belief in *species hierarchy* ("man" at the top), the belief in *intraspecies hierarchy* (some humans are better than others), and the belief in *human improvement and perfectibility* (the Perfectibility of Man).

The thesis of this essay is that this ideology of human progress is very much alive today in the new reproductive technologies. It is important to pursue the connections between such technology and the ideology of human progress for two reasons. First, this underlying motivation has been largely neglected in most of the ethical, legal, political, and scientific discussions that have been generated about this new research in the past decade or more. The arguments center on "*Should* we?"—on limits, guidelines and possibilities—but rarely on "Why do we *want* to?" The second, more important, reason is that now we *can*. A sophisticated and rapidly moving science and technology today enables us to implement this ideology in ways not possible in an earlier age. Artificial insemination, amniocentesis, and maternal blood sampling—to name three techniques now widely used among human beings—make it possible to practice a form of selective breeding. Other techniques have been and are being developed that can allow us to create an ideal, or more nearly perfect, human being. Among the reasons proposed for pursuing this technology are scientific progress, exploring intellectual frontiers, the profit motive, and advancing human liberation, a claim countered by charges of racial, class, and sexual oppression. But the critical factor of the ideology of human progress as yet another powerful motivating force in pursuing reproductive technologies is only occasionally suggested and has not been systematically explored. If we are to understand more about the development of these technologies and the normative issues that surround them, we need to understand the role that the ideology of human progress plays in motivating the pursuit of this research. To do so can supply an important dimension to the implications of this new technology for us and the world in which we live.

Two notes of caution. First, there is no claim here that the ideology of human progress is the only, or even the most important, motivating factor for the pursuit of such science and technology, past or present. Material factors and other ideologies—such as faith in science and technology—are at work and are intertwined. My point, rather, is that the ideology of human progress as a critical motivating force has not been properly explored.[1] Once this ideology is singled out and examined, future work could then show how the ideology and various factors interact to underpin the pursuit of these reproductive technologies.

Second, this essay does not address the question, *should we* pursue these technologies, although the inquiry is motivated by normative concerns about the directions of such research. Rather, I ask "Why *do* we?" The *why* takes on special significance because we *can* now "weed out" the "defectives" and "unfit" *before* they are born, because it is now in the realm of possibility that we could attempt to create a "superman" that a Shaw, Haldane, or Muller merely speculated or fantasized about.[2] In contrast to the pre-molecular biology and pre–recombinant DNA age, we now have—or are pursuing—the scientific and technological means not only to alter the reproductive process (conception outside the womb, gestation in alternative wombs), but also to alter *Homo sapiens* itself (prenatal genetic screening, fetal surgery). Who is this creature we are changing and "improving?" Where does she or he stand in relation to other species, past, present, and future, and to the rest of the natural world? Whether we acknowledge them or not, our beliefs about human nature (which inform our political, social, and economic theories as well as our scientific theories),[3] and about our place in the scheme of things cannot but inform our research and practice in the Gene Age.

The ideology of human progress, which came into its full flowering in post-Darwinian evolutionary theory and the eugenics movement, is a key component of the Western belief system about human nature, about who and what *Homo sapiens* is. Further, and more important, it is key to understanding the ideal that dominant Western thought says human beings *should* be. The difference between today and the past is that we can now act to implement the *should*; we can now pursue the "perfect man." But who is that "perfect man," how have we arrived at this ideal, and how might we implement it? An examination of the role of the ideology of human progress in our biological research should give us a clue.

The first section traces the historical origins of the ideology of human progress and the development of its relationship to evolutionary theory and eugenics in the nineteenth and early twentieth centuries. The next section shows how the ideology was kept alive in the interwar and early postwar periods. With this historical framework in place, the essay moves to the present to explore how the ideology informs reproductive technologies today. For this analysis, the next two sections draw mainly on the new technologies themselves, rather than on writings and testimonies of scientists and researchers, as in the earlier part of the essay. The technologies discussed are limited to those that in some way change the method, process, or outcome of human reproduction; thus, they will include technologies of conception and gestation, prenatal testing, and surgery or other means to change the fetus, genetically or otherwise, before birth. The essay seeks to show the critical interplay of technology and ideology, as the ideology of human progress supports the technology, and the technology in turn modifies and transforms the ideology.

THE IDEOLOGY OF HUMAN PROGRESS: HISTORICAL BACKGROUND

The ideology of human progress consists of three elements: species hierarchy, intraspecies hierarchy, and human improvement and perfectibility. *Species hierarchy* is the belief that humans are the highest order of beings on the planet, if not in the universe. As such, species hierarchy draws on a long tradition of Western religious and philosophical thought that predates the emergence of the ideology of progress as a dominant paradigm and the incorporation of species hierarchy into that paradigm in the nineteenth century. Even within the organic, balanced world view in classical antiquity, the Greek Sophists held that "Man is the measure of all things."[4] Judaeo-Christian Creation ranks man[5] lower than God or the angels, but above the beasts of the field, setting man to rule over them, as well as over woman and the earth.[6] The belief in species hierarchy thus also carries with it a separation, even alienation, from nature (meaning all nonhuman life), and the idea of controlling nature that surfaced before the Scientific Revolution sanctioned and facilitated her demise.[7] The belief often carried with it as well a denial of man's biological nature, or at least a relegating of the body to a lower level, ranking mind over matter, and thus man over woman.

This elevation of the mind (reason or intelligence) over the body (the physical) links species hierarchy with *intraspecies hierarchy,* the belief that some groups of human beings are superior, or inferior, to others. If humans are the highest form of life, what sets them apart from the lower orders? Especially in the Western philosophical traditions, "intelligence," the ability to think or to reason, usually ranks first on the list. Intelligence, reason—this unique ability—it is argued, has enabled humans to fashion language, tools, civilization and culture, setting human society above the level of the animal, the savage, the barbarian. The rational is found superior to the animal, the mind superior to the body. Whether it was an Aristotle "proving" that women and slaves were inferior by nature because of diminished reason and being closer to an animal state, or the eighteenth-century biologists "proving" that Negroes' skulls and physiognomy most closely resembled those of apes,[8] the hierarchical view of human superiority over other species readily translates into support for hierarchy among humans. Superiority is judged by the degree to which people approach the human ideal of intelligence and rationality, inferiority by the degree to which they fall away from that standard, and approach the animal, and thus an uncivilized, state.

Further, intelligence and reason do not stand still. Human beings are capable of improvement, of reaching toward perfection or some ideal. Operating within the confines of prevailing views of human hierarchies of its time, the idea of human perfectibility also antedates the nineteenth-century concept of progress. We find the idea in the teleology of Plato's concept of human nature or of Aristotle's idea of "becoming." In Christian theology, the state of perfection meant regaining the lost Eden, to be attained only after death when, it should be noted, the soul would be separated from the body. When the eighteenth century began the process of joining human perfectibility to the ideology of progress, as Carl Becker has shown, the French *philosophes* transported the "heavenly city" to earth, converting the Perfectibility of Man into a secular faith.[9]

Thus, the ideology of human progress, as it began to emerge before Comte and Spencer, took the following shape. The starting point was the belief that humans are superior to all other living things. It moved on to the idea that some humans are superior to others, a belief based largely on the degree to which they possessed what distinguished and distanced them from animals— namely, reason and intelligence, and, by extension, "civilization" and "culture." Finally, it held that these beings, superior to the animals as well as to certain categories of humans, are capable of improvement and of attaining, or at least aspiring to, an ideal of perfection. The word *ideology,* however, cannot properly be applied to this cluster of beliefs until we review briefly how they became wedded to the belief in progress as it emerged as a dominant ideology in the nineteenth century and how, then, *human progress* became a major ideological underpinning of evolutionary and eugenic thought.

The story of the development of the idea of progress as a central article of the faith of the nineteenth century—capturing alike the industrial entrepreneurs, their socialist critics, and the scientific and intellectual elites—has been told many times over, starting with Bury's now classic account two-thirds of a century ago.[10] Turgot's and Condorcet's view of history as the realization of human progress through reason and science paved the way for the blending of science and progress that was to mark the theories of social development of a Saint-Simon, Comte, and Spencer.[11] Evolutionary thought particularly lent itself to the ideology of progress.[12] In tracing the "tremendous revolution in human thought" from "static creationism" to "evolutionary views of nature" in the period from the late seventeenth to the late nineteenth centuries, historian John Greene's work reveals that all three elements of the idea of human progress existed in pre-Darwinian evolutionary thought. Even before Lamarck in the early decades of the nineteenth century fathered the concept of linear progress that placed humans at the top of the evolutionary hierarchy, Buffon in the mid-eighteenth century had assigned to "man . . . a unique place in the self-sustaining system of nature," his resemblance to the "higher apes" confined to physical structure. An immense separation was created by thought and speech, a God-given gift that gave man the right to rule "over the brute creation" and to seek control over nature for human convenience.[13]

Although Buffon contributed to the development of evolutionary thinking, his view of change in the natural world, according to Greene, was "more in terms of degeneration than of improvement."[14] It was Charles Darwin's grandfather, Erasmus Darwin, in the 1790s who made the leap to link biological improvement to the divine plan, providing a biological basis for the *philosophes'* secularization of the gospel of human improvement and perfectibility. Anticipating Lamarck's idea of organic transformation, Erasmus Darwin wrote in *Zoonomia* in 1794 not only that "all animals undergo perpetual transformations," but also that "all nature exists in a perpetual state of improvement."[15] Attributing this process of natural change to "THE GREAT ARCHITECT! THE CAUSE OF CAUSES!," Darwin, according to Greene, "at one stroke . . . wrested from the new biology a cosmic underwriting of the gospel of human progress and invited Christians to exchange the hope of salvation in the next world for a share in building the increasingly better life on earth."[16]

The idea of human improvement and perfectibility—although not necessarily in a progressive context—also informed the late eighteenth- and early nineteenth-century views of racial differences among humans. Whether holding the racial varieties of human beings as representing different species (polygenists), or as belonging to a single species (monogenists), naturalists and philosophers of this age for the most part saw white Europeans as the highest human form and other races, especially the Negro, as the lowest. Among the monogenists whose views, unlike those of the polygenists, contributed to the development of evolutionary thought,[17] one finds this racial hierarchy, whether in Blumenbach's theory of degeneracy from the white Caucasian to the yellow, brown, and finally black Negro,[18] or in Prichard's theory of progress from the "black races of men" toward the perfection of white Europeans.[19]

Thus, biological connections with beliefs in the superiority of humans in the evolutionary chain, in human hierarchy, and in human improvement and perfectibility were already in place when Charles Darwin's *The Origin of Species* came upon the scene in 1859. The ideology of human progress had begun to forge its links to evolutionary thought. The extent to which Darwin himself forged these connections is a matter of controversy. Darwin saw evolution as a many-branched bush and, in *The Descent of Man* (1871), described a fundamental human biological continuity with the rest of the natural world.[20] Yet he subscribed to the idea of progress of his age and its reigning beliefs in racial, sexual, and intellectual superiority.[21] Whether subsequently distorted or not, Darwin's theories, especially that of natural selection, provided a basis for Social Darwinist and eugenic interpretations.

To the Social Darwinists, reaching back to Lamarck to propose what Mary Midgley has called the "Escalator Fallacy," evolution was "a steady, linear upward movement, a single inexorable process of improvement" from primeval gas to genius and beyond.[22] Humans were not only superior, but also separate from and above nature. Herbert Spencer undertook to transpose *natural selection* into "survival of the fittest,"[23] giving rise to the Social Darwinist, and then the eugenicist, theory of social and moral selection that equated "fittest" with "best." The "fittest" turned out to be those who survived the competition of the economic, social, and intellectual marketplace, proving thereby their biological superiority. Their wealth, position, and accomplishments attested to their superior ability to adapt. Conversely, the losers were the least fit, proving their biological—later interpreted as their genetic—inferiority, and lack of adaptive qualities. As Social Darwinist ideology increasingly provided the climate for the development of eugenic thought, the belief spread that such qualities as temperament and behavior were inheritable, along with physical characteristics.[24] By the "law" of improvement of Spencerian evolution, the fit would be engaged in naturally selecting to improve and perfect the human race, while the unfit would be dying by the wayside unable to adapt. In late nineteenth- and early twentieth-century Britain and the United States—where the ideology of human progress was strongest—according to the eugenicists' own definitions of "fit" and "unfit," such natural selection, much to their chagrin, was not taking place.

The eugenicists' concept of the "best" and "fittest" constituted the ideal or the Perfect Man, to which the human race should aspire. This ideal, particularly as formulated in the United States, was a white male of Northern European stock—a "WASPM" (white, Anglo-Saxon Protestant male). A professional or intellectual from the middle and upper classes, the Perfect

Man was characterized, above all, by *intelligence*. He was the man of science, the expert. He had a superior mind, with even a larger brain than lesser mortals.[25] He was among the worthy, as merit and intelligence were equated. By the same token, the "unfit" were so by nature, passing on their criminality or other bad habits to their progeny.[26] Deficient in the qualities that characterized the fit, the unfit or "defectives"[27] included non-Caucasians and eastern and southern Europeans (especially in the United States), women (some of these obviously being more defective than others),[28] the poor and lower classes generally (especially in Britain), the "criminal types," and other assorted outcasts such as the diseased, the maimed, and the mentally deficient.[29] Eugenicists variously advocated *positive eugenics,* urging the best to reproduce, and *negative eugenics,* preventing those classed as unfit from multiplying.

Thus, in the late nineteenth and early twentieth centuries, the belief in human progress had been fashioned into a powerful ideology through uniting with scientific theories of the day to support particular directions for both scientific thought and practice. The ways in which science was marshaled to justify and promote racist, sexist, classist, and ethnocentric policies and practices have been amply documented.[30] Many actions initiated in the name of science, such as classifications of mental defectives, sterilization laws, IQ tests, and restrictive immigration policies, profoundly affected people's lives. However, the state of scientific knowledge and the level of technological development of the time placed severe limits on the degree to which human reproduction could be directly affected.

In the approximately forty-year period (ending in the 1930s) of open support for the evolutionary and eugenic theories through which the ideology of human progress was articulated, the level of biological research and technological procedures could do little more than facilitate a crude negative eugenics, mainly by preventing certain groups of people from reproducing. Sterilization was the major means employed: castration or vasectomies, practiced mainly on male prisoners, especially sex offenders, and severing or tying the fallopian tubes of women, particularly the "feeble-minded." The first sterilization laws in the United States were enacted in 1907. The Supreme Court upheld a Virginia statue in 1927,[31] and by the end of the 1920s there were sterilization laws in twenty-four states.[32] Other technological means available at that time were abortion and birth control. Abortion, however, was opposed by mainline eugenicists—as those advocating negative eugenics were known—and the practice continued to be illegal throughout this period. While Margaret Sanger's birth control efforts gained support as she became increasingly connected with the eugenics movement, use of birth control technology remained much more a practice among middle and upper class women than among the poor and working class whose birth rate eugenicists wished to limit. Wholesale killing of undesirables—whether in the old-fashioned way of lynching blacks in the South or in the technologically innovative way of the Holocaust, which permanently gave eugenics a bad name—overstepped the line of what was morally and politically acceptable. Thus, sterilization was virtually the only practical, *technological* means available for interfering with human reproduction to carry out eugenic goals in the period prior to World War II. Although the practice cannot be dismissed lightly for its effects in individual, human terms, it was extraordinarily limited as a vehicle to further intraspecies hierarchy or "improvement" of the human race.[33]

Given the state of science and technology of the time, the elimination of "defectives" could only be crudely attempted, and the fashioning of the Perfect Man was still only the stuff of science fiction. The developments in genetic and reproductive research and technology after World War II dramatically changed these possibilities, but not the ideology or the goals.

THE IDEOLOGY OF HUMAN PROGRESS: CONTINUITY AND CHANGE ENTERING THE GENE AGE

As much of society recoiled in horror at the excesses to which ideas about creating superior races and eliminating the inferior had led, the ideology that supported such thought and action presumably would have been laid to rest. Such has not been the case. The belief system that had incubated and found expression in a monstrous offspring was still alive and well, though perhaps altered in outlines and emphasis. Building on the Escalator Fallacy,[34] in the postwar period belief persisted in a linear notion of evolutionary progress: the white Western intellectual (often scientific) male at the top, using his superior mind to control matter and press on to species improvement and perfectibility.

The views expressed at the Darwin centennial celebration held at the University of Chicago in 1959 illustrate how the ideology was carried on. Close to fifty eminent scientists, constituting a "Who's Who" of evolutionary thinkers in the inorganic sciences, the life-sciences, and the human sciences, gathered for five days to present papers and engage in discussion on the theme, "Evolution after Darwin." For the most part, the participants' major work had begun and often been completed in the period prior to World War II. Their thinking, therefore, becomes instructive because they can be seen as transitional figures, trained and working in an earlier intellectual and scientific climate, but living through and meeting just as science was entering the Gene Age.[35] Obviously dissociating themselves from the faulty biology, social analysis, and racially hierarchal excesses of the Social Darwinists and the eugenicists, these scientists nevertheless held fast to the idea of a psychosocial, or mental and cultural, evolution among humans, as distinct from animals, and to the superiority of *mind*.

Sir Julian Huxley, in the convocation address delivered on the final day of the celebration, perhaps best sums up this cast of thought. Entitled "The Evolutionary Vision," his talk focused on the way *mind* had made man the dominant species of the evolutionary process and how mind must create new organizations of thought to meet the massive problems of the age.[36] Although stressing the responsible role man must take for the "right use" of mind, Huxley held that "man's destiny is to be the sole agent for the future evolution of this planet."[37] Evolving as a dominant type through the biological principle of "successive patterns" of "improved organization," man gained and continued dominance through "mind," creating "a succession of successful idea-systems" that prevailed.[38] Human evolution is psychosocial, rather than biological, "achieved by breakthroughs to new dominant patterns of mental organization, of knowledge, ideas, and beliefs."[39] Noting the need to recognize that humans are not born equal, he pointed out that "human progress stems largely from the very fact of their inequality."[40] The "evolutionary truth," concluded Huxley, "shows us our destiny and our duty. It shows us mind enthroned above matter, quantity subordinate to quality. . . . it gives man a potent incentive for fulfilling his evolutionary role in the universe."[41]

Huxley had long since repudiated mainline eugenics, joining with other scientists in the period between the wars to expose the fallacies of its pseudoscience and then to condemn Nazi practices.[42] But, as revealed by his remarks at the Darwin centennial decades later, Huxley had not jettisoned the central tenets of the ideology of human progress—species hierarchy, intraspecies hierarchy, and human improvement and perfectibility—that had underwritten the theories he had rejected and opposed.

The "perfect human being" image of this period as well as of the 1970s and 1980s bears some striking resemblances to the imagined ideal of the 1920s and 1930s: a being of superior intelligence, the product of selective intellectual and cultural evolution who would carry the species to new heights of scientific and cultural advance. Scientists themselves in the earlier period engaged in utopian speculation, portraying futures in which a scientific elite directed positive eugenics to improve and perfect humanity and society. The British biologist J. B. S. Haldane, for example, in his book *Daedalus* published in 1924, projected a world 150 years hence in which ectogenesis was close to universal—only 30 percent of children were still born of woman—and reproduction was "completely separated from [hetero]sexual love." Creating and directing the process was the scientist, "the lonely figure of Daedalus, garbed in black robes," who successfully challenged the gods.[43]

A decade later the American geneticist Hermann J. Muller offered a less fanciful means to perfect the human race through the selective, controlled use of artificial insemination. In *Out of the Night: A Biologist's Views of the Future*,[44] he wrote that " 'in the course of a paltry century or two . . . it would be possible for the majority of the population to become of the innate quality of such men as Lenin, Newton, Leonardo, Pasteur, Beethoven, Omar Khayyam, Pushkin, Sun Yat-sen (I purposely mention men of different fields and races), or even to possess their varied faculties combined.' "[45] Notably absent from his list of the supremely gifted were, among others, blacks (his parenthesis notwithstanding) and women of any race. In further elaborating on such proposals for positive eugenics at the Darwin centennial in Chicago two decades later in 1959, Muller repeated his earlier contention that superior intelligence must be combined with cooperation and an altruistic spirit for the highest phase of human evolution to be attained.[46]

Perhaps the most scientifically rationalized conception of the future human ideal proposed in the period was by J. D. Bernal, who was a physicist rather than a life scientist. In his essay, *The World, the Flesh and the Devil: An Inquiry into the Future of the Three Enemies of the Rational Soul*, originally published in 1929,[47] Bernal's ideal is a new man freed from the rational soul's three enemies—material scarcity (the world), the physical body (the flesh), and human desires (the devil)—through science. Emerging as the rational soul when released finally from the last two constraints, future man is a kind of "pure mind," no longer a human being at all. Rid of his body after passing through the " 'larval' " stage, his brain kept functioning in a shock-proof metal container, his emotions under rational control, Bernal's new man manipulates the environment, mechanizing it, manufacturing life, and improving upon it. Achieving the ultimate in human progress, Bernal's future man will transcend evolution, gaining immortality.[48]

The perfect human being thus fantasized in the 1920s and 1930s reflects values that are classist and elitist, ethnocentric and racist, and patriarchal, building them into the belief in the Perfectibility of Man. The emphasis on intelligence, on a superior brain, effectively separates out the "better sort" through a combination of a class-based educational system and of testing and measuring methods devised by the reigning elites themselves. Despite Muller's nod to a more culturally and racially diverse elite, the kind of scientific and cultural knowledge possessed and advanced by the perfected humans of the future represents what we usually term "Western Civilization," knowledge developed and perpetuated by a largely white, Northern European and American culture. It is a patriarchal vision as well, projecting a male-defined culture, and male-designed means for human reproduction. Not only are women absent from the new elite, women are also devalued. In Haldane's future, women's ovaries are removed and kept alive for a period of twenty years, producing eggs monthly that are fertilized and then gestated in artificial wombs.[49] Muller's reproduction line, still using intact women for his artificial insemination plan known as "eutelegenesis," relegates women's role to "little more than that of conceptual vessels for the sperm of admirable men."[50] Bernal's vision, as Easlea has trenchantly analyzed,[51] is that of a completely masculinized future. Women will either continue to exist as women among the majority of humankind who have not progressed to the advanced rationalist state (attainment of the rational soul is only for the few), or they will become masculinized, joining the disembodied, mentally directed, emotion-free "minority of progressive males" in a "masculinized cosmos. Male mind and intellect will triumph over female matter and emotion, indeed the mind of the future will direct the matter of the entire cosmos."[52] Significantly, included in such direction will be the ability of the new males to create new organisms and manufacture life. The usurping of women's role in reproduction illustrated by these examples represents the ultimate stage of the patriarchal goal to devalue and control women.

Today, the ideology has shifted emphasis. Species hierarchy is assumed, but with perhaps less arrogance than in its optimistic heyday. In light of the nuclear threat, pollution and environmental issues, and recognition of the increased power potentials of science and technology, the belief that humans are earth's highest, most intelligent species tends to take on a cautionary tone. The belief is tempered by a degree of humility and sense of respect for the role humans have had in creating problems as well as benefits: human intelligence, which separates us from the lower animals and which has brought us this great knowledge and power, must be used wisely and responsibly, if the human race is to survive. Similarly, blatant racism and even sexism have become unfashionable—Jensen and Shockley notwithstanding—even as beliefs in human hierarchies persist. Intelligence—as defined by white Western, especially scientific and technological, male "experts"—and the brand of culture it has created remain an effective measure of the ranking system that continues to allow some groups to label and dominate others. Finally, the idea of human improvement and perfectibility, still an article of faith, has undergone a shift. At a time when *perfection* could only be a far-off, imagined ideal, *improvement* loomed as the more important scientific goal and practice. But, as reproductive and biotechnologies have rapidly developed, and fashioning the "perfect human being" comes closer to the realm of the possible, the Perfectibility of Man as a belief has begun to take center stage. Today, improving human life through science and technology is an accepted premise and goal. That such improvement is toward some ideal of perfection is also assumed. But that goal of perfection, while increasingly important, remains largely unspecified, and rarely is it acknowledged or its implications explored.

As in the earlier part of the century, the science and technology of the current age acts to redefine the ideology of human progress, even as the ideology supports and motivates that scientific and technological thinking and practice. In the postwar decades such redefinition has meant a stronger emphasis on the idea of human perfectibility, as reproductive and biotechnology might fashion and implement it. The belief in the Perfectibility of Man will therefore receive greater emphasis here, as current reproductive and genetic technologies are examined for their links to the ideology of human progress. As the technologies themselves, rather than recorded expression of these beliefs, are focused on in the next section, the method of illustration shifts from written and oral testimony of participants to inference from practice. Further inquiry will seek to join the testimony and experience of research participants to the current analysis.

THE NEW TECHNOLOGIES

Three categories of relevant reproductive technology and biotechnology research can be distinguished, even though they will overlap in the discussion that follows. The first covers technologies that affect the *method of reproduction*. This includes processes, from artificial insemination to ectogenesis, that in some way alter what may be called the "normal" method of human conception and gestation: through heterosexual intercourse and full gestation in the womb of the woman in whom conception had taken place. In-vitro fertilization is one example. Not included in this category are techniques that are used at the time of birth, such as fetal monitoring, which may be considered invasive or "unnatural" but do not change how human beings reproduce. The second area, *prenatal testing* or *screening,* includes processes that can detect various chromosomal or developmental anomalies or hereditary conditions, as well as the sex of the fetus. Amniocentesis is a prime example in this category. The third area is *fetal alteration* through surgery or other techniques designed to correct or otherwise change characteristics in the developing fetus. Examples of such procedures are draining fluids and correcting a genetic flaw. Not included in this category are genetic research or practice to cure or treat disease in other than the reproductive setting—for example, genetic engineering applied to a human being after birth, or as used in agriculture or industry. My focus is limited to research and techniques directly affecting human reproduction. In the discussion that follows, distinctions between science and technology will inevitably be blurred because of the ways research, technique, and practice interact when dealing with reproduction.

Who or what is the "perfect human being" as revealed by current reproductive technology and research? Today's technologies tend to define the ideal by presenting a picture of its opposite. That is, while selecting "desirable" genetic traits to create the "perfect child" is still not possible, current technology does enable us to select out growing numbers of "undesirable" characteristics and prevent their immediate reproduction—or in some cases treat them. We will start with procedures for prenatal screening showing how they define who or what is *not* perfect, who or what is "defective," and who or what is devalued and expendable.

In the late 1960s, the procedure known as amniocentesis was perfected for fetal testing and soon became widespread. During the fifteenth to seventeenth week of pregnancy, a small amount of amniotic fluid is withdrawn by a needle from the amniotic sac surrounding the fetus. Testing for various conditions, most notably Down's syndrome and neural tube disorders such as spina bifida, is then possible by culturing the cells thus obtained or by examining the fluid itself. In cases where one or both parents are known or likely carriers of inheritable conditions, such as Tay-Sachs disease, sickle-cell anemia, or beta thalassemia, the amniotic material will be tested specifically for these conditions. The procedure also reveals the sex of the fetus. Amniocentesis is now routinely accompanied by ultrasound or sonography, also a diagnostic procedure, whereby an image of the entire fetus can be projected on a screen, revealing tissue, organ, and skeletal development. Test results from amniocentesis take up to four weeks or longer, placing abortion, if chosen, at the twentieth week, or halfway through the pregnancy (a full-term pregnancy is forty weeks), or possibly later in the second trimester (twenty-four to twenty-six weeks). A newer testing procedure, chorionic villus biopsy or sampling, (CVS) can be performed at the eighth or ninth week, less than two months into the pregnancy. Obtaining the cell sample from the uterus by entering through the vagina and cervix rather than through the abdominal wall, the procedure involves removing tissue from the villi of the chorion membrane surrounding the embryo to test for chromosomal or enzymatic anomalies, similar to testing in amniocentesis. Although CVS can reveal Down's syndrome as well as the sex of the fetus, it cannot fully replace amniocentesis because it does not work for late-developing conditions, such as "neural tube defects . . . some forms of hydrocephalus, some kidney disorders, and some congenital heart problems."[53] An alpha-fetoprotein (AFP) test, however, can now be made on a maternal blood sample as early as nine to twelve weeks into the pregnancy to detect neural tube disorders, that is, conditions in which the neural tube fails to close, such as spina bifida or anencephaly. Because results may be inaccurate or otherwise unreliable at early stages of testing, amniocentesis is a "backup" procedure.[54] The continuing trend of prenatal screening research is to develop reliable procedures that can be performed earlier in the pregnancy, and that reduce the invasiveness and risks still associated with practices in current use.[55]

A study published in the *American Journal of Obstetrics and Gynecology* in 1981 compiled a list of some two hundred conditions that could be diagnosed through prenatal testing.[56] Rapidly developing research in this area continues to expand this list. All diagnosable conditions, however, are not equally life-threatening or severe, nor are all, or even most, conditions routinely tested for. While some of these conditions are "surgically correctable" or "medically manageable" after birth,[57] many others, particularly the most severe, are not. Diagnosis continues to outpace treatment and cure.[58] The focus of prenatal screening remains on diagnosing the most serious conditions, those that will result in immediate or early (often painful) death of the newborn or in severe mental or physical impairment, such as Tay-

Sachs disease, Down's syndrome, or certain neural tube disorders. Because there is now no known cure for such conditions, the option if test results are positive (meaning that something is wrong) is whether to terminate the pregnancy. A number of factors can inform a woman's decision. Of interest in this regard is the role of socially constructed beliefs about human perfection and imperfection and how the technologies themselves inform such beliefs. Included here are issues raised by disabled persons and by the health hierarchy implied in viewing prenatal conditions as pathologies.

Disabled persons have themselves pointed out that all conditions that can be determined through prenatal screening are not equally severe. In most instances, however, the level of severity is not evident until after birth. Some Down's syndrome children are only mildly retarded and impaired; some neural tube defects are less pronounced and are amenable to surgical improvement after birth. It is argued that such persons do not need extraordinary care, do not require institutionalization, and can lead relatively autonomous and productive lives. Marsha Saxton, who was born with spina bifida, has written of suffering greater pain from the pitying and condescending reactions of others to the braces she wore as a teenager, than from the disability itself.[59] Anne Finger, disabled by polio when a young child, has put the point more strongly. She finds in the negative attitudes of society a fear and horror of the "deformed."[60] In an earlier, less sophisticated age, the severely maimed and retarded who survived could be locked away like criminals and drunkards, either legally or in the attic, conveniently hidden from view. In the current climate of greater sensitivity to human rights, an era in which science and technology have allowed more of those once labeled defective to survive, such persons are more visible. As the disabled attest, greater visibility has not necessarily transformed the perfection ideology. Distaste for and rejection of the mentally and physically impaired remain widespread in our society, however masked or modified by the sensibilities of the age. Paradoxically, the science and technology that can provide the means to repair what is defective, can also provide the knowledge, through prenatal screening processes, to enable us to permanently remove "unfortunates" from view.[61] Whether it is "saving" babies who would not survive, or providing criteria on the basis of which survival choices are made, the new technology is in effect helping to formulate new standards of human perfection and imperfection. In this manner, these procedures can force us to confront what the terms *disability, imperfection,* or *defective* mean, as well as the whole belief in human perfectibility.

In the case of prenatal diagnosis, the technology, building on a perfectibility ideology already in place, specifically reshapes those beliefs through the health hierarchy it creates. In the language of prenatal screening, what is imperfect becomes anything that can be diagnosed as a pathological condition, from a fatal affliction to a genetic predisposition for a non-threatening, treatable malady. Classifying such conditions also suggests a hierarchical ordering of prenatally detectable birth defects that are added to and classified as more and more conditions can be diagnosed. Although tests for most of these conditions are not routinely performed—indeed by no means are all women screened[62]—this growing list of diagnosable defects helps to create and reinforce a mode of thought that demands a "perfect child." When the technology was still not available to diagnose such prenatal conditions, even where heritable tendencies were known, a woman could do little more, short of not risking pregnancy at all, than stay healthy and hope for a "healthy child." In a matter of two decades, scientific and technological knowledge has so changed what is possible that concern over having a *healthy child* gives way to pressures to have the *perfect child*. Not only are women urged to undergo amniocentesis at younger and younger ages,[63] they are somehow blamed if they refuse the procedure and then bear a defective child, *as defined by the new technological knowledge*. As Rothman points out, "technology established standards" for the "improved product."[64] There is only the illusion of choice as the structuring and grading of such standards occurs outside of women's control.[65]

One of the conditions on the Stephenson and Weaver list that can be diagnosed prenatally is fetal sex.[66] Any procedure that reveals the chromosomal cell structure will also reveal whether the sex chromosomes are "XX" or "XY" (as well as an anomaly) and thus whether the fetus is female or male. This applies to amniocentesis, as well as to CVS. Aside from its use for known sex-linked heritable conditions, such as hemophilia, obtaining such information serves no particular medical purpose: its revelation is merely a byproduct of the cell-testing process. With the exceptions of India and China where, for different reasons, tests are done for fetal sex, and females are generally aborted,[67] prenatal screening *solely* for this purpose is not prevalent elsewhere and tends to be discouraged. However, prospective parents do want to know the sex, and pregnancies are terminated because the child will be the "wrong sex."[68] As cross-cultural studies have shown, there appears to be a universal—though not unanimous—preference for sons, particularly as the first-born.[69] Although there are clear economic reasons for preferring sons in agricultural societies, such preferences also reflect the prevalence of a patriarchal culture in which males are valued over females, even by women. This is a case where technology does not set the standards, but where technology makes it possible to implement other standards—in this case patriarchal values—that are socially constructed. Thus, sex—or, more precisely, gender[70]—is a prenatal "defect" in the sense that it, like other standards of perfection, is socially constructed.[71] To the extent that being female has less value than being male, a diagnosis of female fetal sex joins the list of diagnosable mental and physical conditions that could be termed imperfections by virtue of today's reproductive technology making it possible to define them as such. Sex selection prior to conception, when perfected, will further demonstrate how patriarchal values (as well as class and race issues) intersect with the new technology and the ideology of human perfectibility.

Some of the reproductive technologies now practiced, or at least possible, begin to frame a picture of human perfection as well as imperfection, even as they provide the means for implementing that image. Artificial insemination is a case in point. An old technology, it began to be used to breed farm animals at about the turn of this century; with the freezing of sperm perfected in the mid-1960s, such use became widespread.[72] Although its use among humans had been restricted, artificial insemination has increased recently as a result, in part, of changes in the climate of thought and of new technologies.

One new use, although the demand is apparently not overwhelming, has come with the establishment in the 1970s of the Repository for Germinal Choice in Escondido, California. Quickly dubbed the "Nobel Sperm Bank" because it sought the sperm of Nobel prize winners in the hope of propagating a race of scientific geniuses,[73] the enterprise was soon subjected to adverse publicity and ridicule. Raising distasteful eugenic connotations, especially with the racist taint from William Shockley's much-publicized sperm donation,[74] the enterprise was on shaky scientific ground, harking back to the outdated, biologically incorrect genetics on which the eugenics movement had been based. Nevertheless, this sperm bank's very existence and its support in some quarters have served to reinforce the belief that intelligence, especially scientific intelligence, is the highest human attribute and the human ideal is a scientific genius, who is white and male. The Repository for Germinal Choice perpetuates the notion that it is only the man's genes that count, the woman serving as vessel to reproduce an improved human race.

Surrogate motherhood, also made possible by artificial insemination, perpetuates this idea as well. The practice has gained favor among heterosexual couples in which the woman is unable to conceive or carry her own child to term. Another woman contracts to be inseminated with the man's sperm, carry the fetus thus conceived, and at birth deliver the baby to the couple. Not the least among the growing controversies surrounding this practice is the argument that the so-called surrogate mother takes on the vessel role.[75]

Another use of artificial insemination that combines specifically with new technology is for sex *pre*selection. Research is going forward on a number of techniques, the claimed success rates varying. These techniques involve separating X- and Y-bearing sperm, for example, through centrifugal or electrical methods, with the desired sperm then injected into the woman's vagina. The process may also involve preparing the woman to be biochemically less or more receptive to each type of sperm.[76] Artificial insemination need not be involved if sex preselection technologies were to be used during fertilization outside the womb.[77] Whether the ideal of a first-born male would be followed if such preselection technologies became routinely available is difficult to predict, but the technology would make it possible and would preclude the moral issues associated with terminating pregnancies.

The process of conception outside the womb, or in-vitro fertilization, marks a dramatic break with the reproductive technologies discussed so far. Until this point, only the male reproductive gametes could be separated entirely from their human donor, both physically and in time, in order to perform their role in fertilization. But, with the conception and birth of Louise Brown, the first "test tube baby," in 1978, the female's egg was now similarly detached from its human host, kept alive, and induced to unite with male sperm. Successful egg retrieval was considerably aided by an earlier developed method of hormonal stimulation to increase ovulation. After the egg is fertilized in the petri dish and reaches a four- to eight-cell stage (in about two or three days), it is transferred to the woman's womb for implantation.[78]

A number of other technologies have rapidly followed the development of this first successful ex-utero reproductive process. Among these are freezing embryos, embryo transfer, and research on freezing the eggs themselves. The "world's first frozen-thawed baby" was born in Australia in March 1984.[79] The first child born as the result of embryo transfer from one woman to another (embryo "flushing") was in California in January 1984.[80] Reproductive research on animals, which usually precedes applications to humans, continues in such areas as embryo transfer from one species to another, and cloning. Although tadpoles (as well as carrots) have been cloned, the reported cloning of mice in 1980 has since been discredited, and to date no adult mammal has been cloned.[81] Except for some discussion of possibly attempting to gestate a human fetus in a sheep or other mammal, experiments to bring a human fetus to term outside a woman's uterus have focused on constructing machinery, an artificial womb, to duplicate the human environment.

THE NEW TECHNOLOGIES AND THE IDEOLOGY OF HUMAN PROGRESS

Some of the ways these technologies and procedures reflect or modify concepts of human perfection and imperfection have been suggested. Let us look more systematically at this interaction in relation to the belief in human perfectibility specifically, and the ideology of human progress generally. Included among the factors behind the intense efforts of science and technology to enable the once infertile to reproduce is an unquestioned assumption of species hierarchy: the human race must go on no matter what extraordinary measures need be undertaken. In-vitro fertilization, for example, is a long, difficult, frustrating, painful, usually unsuccessful, not to say expensive, process, that reflects this belief.[82] The availability and continued pursuit of infertility technologies escalate the demand for such measures and the desire to pass on one's own genes to affirm one's humanity. A belief in species hierarchy also informs the extensive use of animals to further research in human reproduction.

That these technologies are available selectively, making it possible only for *some* people's genes to be passed on, reflects and furthers the belief in intraspecies hierarchy. In a more sensitized social climate today, the interaction of technologies with race, class, and gender issues becomes more complex. For example, when laws were enacted and public programs set up in the United States in the 1970s to screen for the trait for sickle-cell anemia among prospective black parents—one out of twelve blacks will carry the recessive gene—the measures were welcomed as a response to a need and a demand.[83] Yet such screening can be used to discriminate in employment and can be open to charges of genocide, if the purpose appears to be to limit severely the birth of black babies and thus the size of the black population. Such charges have been made. That public programs were needed at all underscored the fact that the availabilities of screening procedures and prenatal care are related to a class-based health care system that is inadequate to the needs of those low on the socioeconomic scale. Race and class factors that have operated to deny large groups of women access to new reproductive technologies—and thus their right to choose as well as refuse them—further reflect the coercive ways technologies are used.[84] Surrogate motherhood is typically class-based, the professional couple contracting with a woman from a working-class background to carry and bear *their* child for a fee.[85] Women from disadvantaged and Third World groups have been and are being used in reproductive research, such as new birth control measures and human egg production and retrieval.[86]

Thus, while the new technologies have not *created* race, class, and gender hierarchies, their development and use *take advantage of* and *support* prevalent beliefs and practices concerning individual and group superiority and inferiority. Their existence can act to reinforce and encourage certain hierarchies. The devaluing of women is a further case in point. When infertility is treated as a pathology and intense efforts to overcome infertility focus especially on women, the view is encouraged that reproduction is a woman's major function. Long made to feel that they are not "real women" unless they can conceive and bear their own child, women fasten their hopes on each new anti-infertility technique as it is discovered, even as the availability of the technique exploits and feeds the mother syndrome. When the surrogate mother serves as vessel to carry another's child, the ancient and medieval notion is recalled of a fully formed infant implanted as a male seed, the woman acting only as temporary incubator. As a woman undergoes procedures such as in-vitro fertilization and embryo transfer, the vessel image is reinforced, along with that of woman as a passive, inert entity in the reproductive process. The new technologies make it increasingly possible to reduce women to ovaries, fallopian tubes, and uterus, just as in the late nineteenth-century "science" sought to "prove" that such was indeed women's true calling and fundamental nature.[87] Ectogenesis, which would substitute machines and mechanical processes for female reproduction, could signify the ultimate devaluation—no need for women at all.

The human perfectibility ideal, while echoing earlier beliefs, takes on new dimensions under the impetus of the new technologies. In the Repository for Germinal Choice, we find the old ideal of the super-intelligent white male scientist still alive and well—if more benign than the visions of a Haldane, Bernal, or Muller. While "superior" notions persist—as fed by currently operative race, class, and gender beliefs—the new technologies tend to frame the ideal in a less threatening, more accessible context. As these reproductive technologies become available and accessible to the middle and upper-middle classes so that parents can be assured of a more nearly perfect child, the individual parents' genes tend to become the criteria on which to fashion the ideal. It is the individual's genes that count. To produce a "better" or "improved" offspring becomes a matter of selecting or rejecting (and in future perhaps perfecting) the genes of the individuals involved. The standards appear to become privatized. Whether you have superior or inferior genes to bequeath to your offspring is a matter of personal responsibility and choice. Although one may argue that reproductive choice belongs at the individual and private level and that the individuals involved (rather than scientists, government, or anyone else) should determine what kind of people we shall be in the future,[88] the belief that we are setting our own standards in this instance is itself an illusion. As I have sought to illustrate in this essay, social criteria combine with the possibilities and limits of the new technologies to set the framework and standards that govern individual choice, including the ideology of individual choice itself. The belief in the power of one's own genes to make an impact, and the belief that they should, become a key element of the ideology of human progress as it is shaped by today's reproductive technology.

The picture that emerges of the ideology of human progress in relation to the new reproductive technologies needs finally to be explored in light of two further areas of scientific and technological inquiry and practice: DNA research and fetal surgery. In 1953, the discovery of the "double helix" revealed the structure of the nucleotides of the DNA, which encode the information in our genes for all life processes.[89] When combined with developments in molecular biology, this discovery has resulted in an explosion of research, much of which profoundly affects work in human reproduction. For example, certain tests used in prenatal screening depend on identifying and locating genetic factors that cause genetic-based heritable conditions. Identifying and locating such genes involves a process known as "mapping." Although only a relatively small number of the up to 100,000 or so genes that make up the entire human gene complement have been mapped so far, high-speed computers are enabling the process to go forward at a rapid rate. It is conceivable that the point will be reached in the foreseeable future when all human genes will have been so mapped. In addition to identifying particular genes, their absence, or their flawed character in a fetus (or a living person), such knowledge can also be employed in recombinant DNA technology, known as "gene-splicing," or "genetic engineering." Bits of genetic material are removed from the DNA of one organism and spliced into the DNA of a microorganism to be cultured, the replicated genetic material then transplanted to a third organism to transform its DNA in some way. The technology has been used particularly in agriculture to make plants disease resistant. In medicine, the splicing and culturing parts of the process have been used to produce human insulin as a substitute for animal insulin to treat diabetes. To be applied to the reproductive process, recombinant DNA technology could involve a further technology—fetal surgery. Operating on the fetus while it is still in the womb, however, is still limited, confined to difficult and risk-laden procedures to correct what would almost surely be a fatal condition. Such procedures would include draining fluids from the fetal brain or other organs, or operating to correct a severe structural impairment.[90]

As techniques improve, if they are pursued, it could become possible to combine fetal surgery with recombinant DNA procedures to correct the genetic makeup of the fetus. Under current guidelines of the National Institutes of Health (NIH), recombinant DNA may be used only for corrective, as opposed to genetically altering, procedures in humans. This means that NIH will review experiments for somatic cell, or corrective, gene therapy, but not those that would alter the genetic makeup of the organism to be passed on to future generations, or germline gene therapy.[91] These restrictions, however, do not apply to animal experiments. Germline gene transplants with recombinant DNA, which must be done in the early embryo, have been performed successfully with mice, and research continues.[92] Perfecting such procedures could bring us closer to accepting their application to the DNA of a human embryo. As the quest goes on to prolong the gestation period outside the womb,[93] the ability to use recombinant DNA for various kinds of corrective manipulations would be greatly enhanced.

Others have raised the specter of a world in which human beings are made to order—with Huxley's *Brave New World* as the prototype[94]—perhaps with more insistence in recent years as the technology begins to border on the possible, if not the probable. Ranging from a vision of a highly repressive, regimented society to less global, yet equally critical, possibilities of the custom-designed child,[95] a dystopian picture emerges.[96] Although promoters of the new research have engaged in some long-range eugenic speculations,[97] defense of the new research tends to stay close to immediate medical and scientific benefits, such as eliminating disease, or relieving pain and suffering, within a vaguely articulated value framework of scientific and technological progress. My purpose here is not to assess these various projections and claims, or to add to "brave new world" alarms. Rather, I seek to expose and explore a key ideological underpinning for such research: the belief in human progress and, particularly, in the Perfectibility of Man.

CONCLUSION

As the research into and the practice of reproductive and biotechnologies continue, ethical, legal, and some political-economic issues are debated. But far less attention is paid to the ideological underpinnings of such research. In particular, the belief in species hierarchy, intraspecies hierarchy, and human improvement and perfectibility—that is, the ideology of human progress—has not received its due. I have argued that this ideology that informed evolutionary and eugenic thinking of the late nineteenth and earlier twentieth centuries informs research and practice in reproductive and biotechnology today. The ideology in operation often supports racist, classist, ethnocentric, and patriarchal values. There is a key difference, however, between the earlier period and now, stemming from the state of scientific and technological research in the two eras. Even in the pre-Hitler climate when eugenic thinking was respectable and some practices could be implemented, technical (as opposed to legal and social) practices to eliminate the "unfit" were confined almost entirely to limited use of sterilization. Genetic research to track down and treat heritable diseases was only just beginning. Improving and perfecting the human race by selecting out and breeding superior specimens remained in the realm of science fiction. The scientific and technological discoveries of the postwar era changed the situation dramatically. Growing genetic knowledge, molecular biology, and medical techniques were beginning to make it possible, for the first time, to alter the *process* and *method* of human reproduction and even the character of the fetus itself. From being able to conceive a child outside the womb, to reimplanting or transplanting an embryo, to screening fetuses for chromosomal, structural, and organic status, to procedures to change the fetus, science and technology are gaining the potential to bring abut major changes in human reproduction. The pursuit and direction of these changes will accord with and implement whatever societal beliefs and visions prevail. As stated at the outset, there is a complex of beliefs that interact to motivate and support such research and its use. This essay has focused on one, which has been neglected: the ideology of human progress. I have sought to show how the prevalence of the ideology is evident in the technology itself, even as the technology transforms the ideology.

Species hierarchy is a given, a holdover not only from certain evolutionary thinking but also from a long religious and philosophical tradition in the West. The way in which reproductive technologies and biotechnology are pursued further supports the belief in human superiority in the evolutionary chain. The belief takes on a solipsistic aspect as more means are found to enable people to pass on their own genes, and as they are encouraged to do so. The selectivity of the research and its application both support and promote the belief in human hierarchies. Benefits as well as disadvantages of the research break along class, race, and cultural lines, and women are systematically devalued. With the improvement and eventual perfection of the human race brought closer to attainment by the new technology, both the human hierarchy and perfectibility aspects of the ideology are transformed. High intelligence, health, and being privileged, white, and male remain among the criteria for the perfect human being. But developments in prenatal screening focus special attention on health, which can now be more precisely defined as the absence of imperfections that can be diagnosed. As the list of diagnosable fetal conditions grows and can be ranked for their severity, the concept of class can be extended to health. An economically individualistic society that holds material well-being to be attainable by all who will use the available opportunities and, coordinately, that poverty is one's own fault, can apply this concept to mental and physical well-being. There is no excuse for mental or physical disability, just as there is no excuse for being poor, it could be argued, because the opportunities to prevent their appearance are available. "Imperfect" human beings then carry the added stigma of being resented for having been born at all, while the mother or parents who allowed this to happen are blamed as well. This kind of thinking, which is an emerging part of the transformed ideology, also ties in with the stronger element of privatization that now obtains as choices are made about the new technologies. Not just health but perfection for one's own child is encouraged. As technology sets the parameters and standards, its ideal of perfection becomes incorporated into individual reproductive goals. Finally,while women may appear to benefit from some of the new technologies,their role as active human beings in the reproductive process is devalued. Technology not only tells them that they must bear "perfect" children. It also suggests that only their eggs are really needed, their wombs "rented" as temporary incubators until a more predictable and controllable vessel can be found. Indeed, improvement of the human race may best be attained by achieving the age-old male dream of procreation without the aid of women at all.

Today's reproductive technologies and biotechnologies strengthen and transform the ideology of human progress by being able to implement these beliefs. This essay has concentrated on the technologies themselves, to delineate this strengthening and transforming process. But we must also look elsewhere to broaden our understanding of the dimensions of this process. For example, we need to explore the views expressed by scientists and other researchers and professionals involved, as well as the reactions and experiences of those affected by or participating in these technologies. We also need to look more critically at beliefs in human superiority and human perfectibility as valid supports for scientific and technological research and practice in human reproduction—and to look for possible alternatives.

Among these might be less species-centered, less hierarchical, and more holistic views both of human nature, and of human beings' place in the cosmos.

NOTES

1. Among the important discussions of the new reproductive technologies in relation to eugenics have been those that focus particularly on the impact on women and on the issue of control. See, for example, Ruth Hubbard, "Eugenics and Prenatal Testing," *International Journal of Health Services* 16, no. 2 (1986): 227–42; Gena Corea, *The Mother Machine: Reproductive Technologies from Artificial Insemination to Artificial Wombs* (New York: Harper & Row, 1985); Robyn Rowland, "Technology and Motherhood: Reproductive Choice Reconsidered," *Signs: Journal of Women in Culture and Society* 12, no. 3 (Spring 1987): 512–28.

2. George Bernard Shaw, *Man and Superman* [1901–3], in *Plays by George Bernard Shaw* (New York: New American Library, 1960); J. B. S. Haldane, *Daedalus, or Science and the Future* (New York: E. P. Dutton, 1924); Hermann J. Muller, *Out of the Night: A Biologist's View of the Future* (New York: Vanguard, 1935). See further discussion of Haldane and Muller below.

3. See Alison Jaggar, *Feminist Politics and Human Nature* (Totowa, N.J.: Rowman & Allanheld, 1983).

4. Protagoras, fifth century, B.C.

5. In this essay, the word *man* or male pronouns will be used only when *male* is meant literally, or when an author referred to follows this usage. In all other instances where human beings are signified, appropriate gender-neutral language will be used.

6. Rosemary Radford Reuther, *New Woman, New Earth: Sexist Ideologies and Human Liberation* (New York: Seabury Press, 1976); Elizabeth Dodson Gray, *Green Paradise Lost* (Wellesley, Mass: Roundtable Press, 1981).

7. See Carolyn Merchant, *The Death of Nature: Women, Ecology, and the Scientific Revolution* (New York: Harper & Row, 1980).

8. John C. Greene, *The Death of Adam: Evolution and Its Impact on Western Thought* (Ames, Iowa: Iowa State University Press, 1959), chap. 6, 8.

9. Carl L. Becker, *The Heavenly City of the Eighteenth-Century Philosophers* (1932; reprint, New Haven, Conn.: Yale University Press, 1959). On the concept of perfectibility in the history of political thought, see Virginia Lewis Muller, *The Idea of Perfectibility* (Lanham, Md.: University Press of America, 1986).

10. J. B. Bury, *The Idea of Progress* (New York: Macmillan, 1923).

11. Krishan Kumar, *Prophecy and Progress: The Sociology of Industrial and Post-Industrial Society* (Harmondsworth, Eng.: Penguin Books, 1978). See especially the clear development of this idea in Kumar's opening chapter.

12. Greene, *Adam;* Kumar, *Prophecy,* p. 18.

13. Greene, *Adam,* pp. 144, 154.

14. Ibid., p. 154.

15. Quoted in ibid., p. 168.

16. Ibid., pp. 168–69.

17. Ibid., p. 222.

18. Ibid., p. 223.

19. Ibid., p. 242.

20. Charles Darwin, *The Descent of Man* (London: John Murray, 1871).

21. Both Greene (*Adam*) and Mary Midgley, *Evolution as a Religion: Strange Hopes and Stranger Fears* (London and New York: Methuen, 1985), point out that

Darwin specifically rejected the evolutionary "ladder" and improvement theories. But see Brian Easlea, *Science and Sexual Oppression* (London: Weidenfeld and Nicolson, 1981) on "Darwin's Two Souls," chap. 5.

22. Midgley, *Evolution*, pp. 6, 34. See also Stephen Jay Gould, *The Mismeasure of Man* (New York: W. W. Norton, 1981), pp. 24, 159.

23. According to Greene (*Adam*, pp. 299–300), Darwin was never happy that he had adopted Spencer's term.

24. Daniel J. Kevles, *In the Name of Eugenics: Genetics and the Uses of Human Heredity* (New York: Alfred A. Knopf, 1985), pp. 70–71.

25. Gould, *Mismeasure*.

26. Kevles, *Eugenics*, pp. 70–71.

27. Originally used only as an adjective, *defective* was first used as a noun in the 1880s. See *Oxford English Dictionary* (New York: Oxford University Press, 1971), 1: 670.

28. Some theorists of this period equated women and Negroes, placing them on the same low level. See Barbara Ehrenreich and Deirdre English, *For Her Own Good: 150 Years of the Experts' Advice to Women* (Garden City, N.Y.: Anchor Press/ Doubleday, 1979), p. 117. Where this might have placed black women presumably was not specified.

29. The American psychologist Henry Goddard classified the mentally deficient further as idiots, imbeciles, and morons, in ascending order. See Kevles, *Eugenics*, pp. 77–78, and Gould, *Mismeasure*, pp. 158–59.

30. See, for example, Gould, *Mismeasure*; Ehrenreich and English, *Her Own Good*; Kevles, *Eugenics*; and Linda Gordon, *Woman's Body, Woman's Right: A Social History of Birth Control in America* (New York: Penguin Books, 1977).

31. *Buck* v. *Bell*, 274 U.S. (1927), 201–3.

32. Kevles, *Eugenics*, pp. 110–11. Legally coerced sterilization was opposed in Britain, eugenicists there favoring "voluntary sterilization."

33. According to Kevles, ibid., p. 112, some twenty thousand legal sterilizations were performed in the United States by the mid-1930s.

34. Midgley, *Evolution*, p. 6.

35. For example, discovery of the DNA structure was published in 1953: James D. Watson and Francis Crick, "Molecular Structure of Nucleic Acids: Structure for Deoxyribose Nucleic Acid," *Nature* 171 (25 April 1953): 737–38.

36. In Sol Tax and Charles Callender, eds., *Issues in Evolution: The University of Chicago Centennial Discussions*, vol., 3, *Evolution after Darwin: The University of Chicago Centennial*, ed. Sol Tax (Chicago: University of Chicago Press, 1960), pp. 249–61. See also Hubbard, "Eugenics," for a 1941 quote from Julian Huxley and further evidence of the persistence of eugenics thinking.

37. Tax and Callender, *Evolution*, p. 252.

38. Ibid., pp. 250–51.

39. Ibid., p. 251.

40. Ibid., p. 258.

41. Ibid, pp. 260–61.

42. Kevles, *Eugenics*, pp. 127–28.

43. As described in Kevles who notes that Aldous Huxley's *Brave New World*, published in 1932, satirized *Daedalus*, upending Haldane's vision into a dystopian nightmare. Ibid., pp. 185–86.

44. Muller, *Out of the Night*.

45. Quoted in Kevles, *Eugenics*, p. 190.

46. In Sol Tax, ed., *The Evolution of Man: Mind, Culture, and Society,* vol. 2, *Evolution after Darwin: The University of Chicago Centennial,* ed. Sol Tax (Chicago: University of Chicago Press, 1960), pp. 234–62. See also Midgley, *Evolution,* p. 34.

47. J. D. Bernal, *The World, the Flesh and the Devil: An Inquiry into the Future of the Three Enemies of the Rational Soul* (1929; reprint London: Cape, 1970).

48. As quoted and discussed in Brian Easlea, *Fathering the Unthinkable: Masculinity, Scientists and the Nuclear Arms Race* (London: Pluto Press, 1983), pp. 150–55.

49. Kevles, *Eugenics,* p. 185.

50. Ibid., pp. 190–91.

51. Easlea, *Fathering*

52. Ibid., p. 154–55.

53. Barbara Katz Rothman, *The Tentative Pregnancy: Prenatal Diagnosis and the Future of Motherhood* (New York: Viking, 1986), p. 221.

54. Ibid., 221–22.

55. For further analysis of key issues surrounding prenatal screening, see Patricia Ann Spitzig, "Genetic Screening Tests: Science, Ethics, and Policy Options," M.S. thesis, Department of Political Science, Massachusetts Institute of Technology, February 1987. I thank Pat Spitzig for generously sharing her research-in-progress with me.

56. Sharon R. Stephenson and David D. Weaver, "Prenatal Diagnosis—A Compilation of Diagnosed Conditions," *American Journal of Obstetrics and Gynecology* 141, no. 3 (1981): 319–43.

57. Ibid.

58. Rothman, *Pregnancy,* pp. 230–33, points out that the use of prenatal screening has been accompanied by markedly reduced activity to seek cures for the most serious diseases, thus suggesting a possible cause-effect linkage to be explored.

59. Marsha Saxton, "Born and Unborn: The Implications of Reproductive Technologies for People with Disabilities," in *Test-Tube Women: What Future for Motherhood?,* ed. Rita Arditti, Renate Duelli Klein, and Shelley Minden (London and Boston: Pandora Press, 1984), pp. 298–312. See also Marsha Saxton, "Prenatal Screening and Discriminatory Attitudes about Disability," *GeneWATCH* 4, no. 1 (January–February 1987): 8–10 (Boston: Committee for Responsible Genetics).

60. Anne Finger, "Claiming All of Our Bodies: Reproductive Rights and Disability," in *Test-Tube Women: What Future for Motherhood?,* ed. Rita Arditti et al., pp. 281–97.

61. I do not suggest an anti-choice position; I support each woman's absolute right to control her body and her reproductive processes. I question, rather, the context in which choices are made, specifically, how the availability and nature of technologies may help to shape the values that structure that choice.

62. Rita Arditti, Renate Duelli Klein, and Shelley Minden, eds., *Test-Tube Women: What Future for Motherhood?* (London and Boston: Pandora Press, 1984).

63. Rothman, *Pregnancy.* Originally recommended only for women forty years and older because of the statistical correlation of age and Down's syndrome, the suggested age for amniocentesis has been dropping steadily to include women in their early thirties, even where the pregnancy is not considered "high risk" for any specific reason.

64. Ibid., p. 227.

65. Barbara Katz Rothman, "The Meanings of Choice in Reproductive Technology," in *Test-Tube Women: What Future for Motherhood?,* ed. Rita Arditti et al., pp. 23–33; and Rothman, *Pregnancy.*

66. Stephenson and Weaver, "Prenatal Diagnosis."

67. Viola Roggencamp, "Abortion of a Special Kind: Male Sex Selection in India," in *Test-Tube Women: What Future for Motherhood?*, ed. Rita Arditti et al., 266–77; and Helen Bequaert Holmes, "Sex Preselection: Eugenics for Everyone?" in *Biomedical Ethics Reviews—1985*, ed., James Humber and Robert Almeder (Clifton, N.J.: The Humana Press, 1985), pp. 39–71.

68. That amniocentesis results are obtained as late as five months into the pregnancy has contributed to discouraging its use for sex selection. However, as CVS becomes more widespread—a test that can be done as early as two months—the use of prenatal screening specifically for sex selection may indeed grow.

69. Jalna Hanmer, "Sex Predetermination, Artificial Insemination and the Maintenance of a Male-dominated Culture," in *Women, Health and Reproduction,* ed. Helen Roberts (London: Routledge and Kegan Paul, 1981), pp. 163–90; Jalna Hanmer, "Reproductive Technology: The Future for Women?" In *Machina Ex Dea: Feminist Perspectives on Technology,* ed. Joan Rothschild (New York and Oxford: Pergamon Press, 1983), pp. 183–97.

70. Feminist analysis makes an important distinction between *sex* and *gender. Sex* is a descriptive category used to designate female and male. *Gender* is a socially constructed category designating characteristics labeled *feminine* and *masculine*.

71. See the nicely developed argument in Rothman, *Pregnancy,* chap. 5, that sex can be classed as a "defect."

72. Hanmer, "Reproductive Technology;" Corea, *Mother Machine*.

73. When Robert Graham, the sperm bank's founder, sought to name it after biologist Hermann Muller who had died in 1967, Muller's widow objected because it stressed breeding for genius alone, rather than for altruism. See Kevles, *Eugenics,* pp. 262–63; see also Corea, *Mother Machine,* pp. 24–26.

74. Kevles, *Eugenics,* pp. 271–72. Other sperm banks, which have been established in growing numbers in recent years, while not explicitly elitist as is Graham's enterprise, can nevertheless lend themselves to supporting eugenics ideology; see Corea, *Mother Machine,* chaps. 2, 3.

75. The "Baby M" case in New Jersey, in which Mary Beth Whitehead sought to keep the child she had borne rather than delivering it to the Sterns, the contracting couple, has focused public attention on surrogacy and its legal, ethical, and feminist issues. See, for example, the discussion by Rita Arditti, "Wombs for Rent, Babies for Sale," *Sojourner* 12, no. 7 (March 1987): 10–11. Although legal opinion is moving against commercial surrogacy—the Whitehead-Stern contract was invalidated by the New Jersey Supreme Court in February 1988—the deeper issues remain.

76. Hanmer, "Reproductive Technology;" Corea, *Mother Machine,* chap. 10; Walter Sullivan, "New Way Devised to Pick Child's Sex," *New York Times,* 23 September 1987, p. D30. For further discussion of sex selection techniques, especially related to eugenics, see Holmes, "Sex Preselection."

77. Although the report in late 1986 by a private fertility clinic in New Orleans of the birth of the first sex-preselected in-vitro fertilization baby (a boy) has been questioned, the possibility exists and is a source of concern. See discussion in Ann Snitow, "The Paradox of Birth Technology," *Ms.* 15, no. 6 (December 1986): 76. Professor Jacques Testart, a French specialist in both in-vitro fertilization and embryo freezing, has called a halt to any such manipulation of the egg or embryo in his future research; see *Hastings Center Report* 16, no. 6 (December 1986): 4.

78. Corea, *Mother Machine,* chap. 7.

79. Ibid., p. 118.

80. Ibid, p. 80. This procedure involves use of a donor egg. The woman donating the egg is inseminated, the resulting embryo then "flushed" from her uterus after five

days and inserted in the uterus of the wife of the man whose sperm was used, his wife then attempting to carry the fetus to term.

81. Hanmer, "Reproductive Technology:" Corea, *Mother Machine*. Duplicating prize cattle, however, has been successfully effected through inserting genetic material from 36-cell embryos into donor eggs, which are then gestated in surrogate cows. See Keith Schneider, "Cloning Offers Factory Precision to the Farm," *New York Times,* 17 February 1988, pp. A1, D6. Recent research has also produced embryonic cloning in which animal embryos have been split to form twins and quadruplets. Embryonic cells from two different species have also been combined to create chimeras. See C. B. Fehilly, S. M. Willadsen, and E. Tucker, "Interspecific Chimaerism Between Sheep and Goat," *Nature* 307: (1984) 634–36.

82. There are a reported 200 IVF clinics throughout the world, 150 of these in the U.S., and an estimated 3,000 babies born so far through in-vitro fertilization. In the United States, costs are $4,000 to $5,000 per attempt. Although success rates as high as 25% have been reported by some clinics, statistics do not always distinguish among such factors as pregnancies achieved, live births, or number of attempts; overall success for live births appears to remain low, less than 10%. See discussion, for example, in Rowland, "Technology."

83. Kevles, *Eugenics,* pp. 255–56.

84. The Hyde Amendment (1976; upheld by U.S. Supreme Court, 1980) prohibits use of federal funds for abortions, making access to abortion difficult for women on public assistance, except in those states allowing state fund use. Yet, even though sterilization without consent is illegal, such women are still sterilized against their will. See Helen Rodriguez-Trias, "Sterilization Abuse," in *Biological Woman—The Convenient Myth,* ed. Ruth Hubbard, Mary Sue Henifin, and Barbara Fried (Cambridge, Mass.: Schenkman, 1982), pp. 147–60.

85. As illustrated in the "Baby M" case: see n. 75 above.

86. Arditti et al., *Test-Tube Women;* Corea, *Mother Machine*.

87. Marian Lowe and Ruth Hubbard, eds., *Woman's Nature: Rationalizations of Inequality* (New York and Oxford: Pergamon Press, 1983); Ehrenreich and English, *Her Own Good*.

88. For a different view, see Jonathan Glover, *What Sort of People Should There Be? Genetic Engineering, Brain Control and Their Impact on Our Future World* (Harmondsworth, Eng.: Penguin Books, 1984).

89. Watson and Crick, "Nucleic Acids."

90. Robin Marantz Henig, "Saving Babies before Birth," *The New York Times Magazine,* 28 February 1982, pp. 18–22, 26, 28, 45–46, 48. In October 1986, the press reported on a "pioneering operation," performed more than a year earlier in July 1985, in which a twenty-three-week-old fetus was removed from the womb, operated on for a blocked urinary tract, and returned to the womb, to be born later by Caesarian section. See Sandra Blakeslee, "Fetus Returned to Womb Following Surgery," *New York Times,* 7 October 1986, pp. C1, C3.

91. See "Points to Consider in the Design and Submission of Human Somatic-Cell Gene Therapy Protocols," available from National Institutes of Health, Rockville, Md; and NIH's "Guidelines for Research Involving Recombinant DNA Molecules," *Federal Register,* 7 May 1986, pp. 16, 958–85. See also Harold M. Schmeck, Jr., "Relaxation Urged in Gene Splicing Rules," *New York Times,* 30 September 1986, p. C9.

92. Two recent reports describe gene transplants in mice, one, to cure a hereditary blood disease and the other, to produce fertility, both using germline gene therapy. See Frank Costantini, Kiran Chada, and Jeanne Magram, "Correction of Murine-Beta-Thalassemia by Gene Transfer into the Germline," *Science* 233 (12 September 1986):

1192–94; and Harold M. Schmeck, Jr. "Gene Transplant Causes Fertility," *New York Times,* 5 December 1986, p. A17.

93. Corea, *Mother Machine,* chap. 12.

94. Aldous Huxley, *Brave New World* (1932; reprint, New York: Bantam Books, 1958).

95. Helen B. Holmes, Betty B. Hoskins, and Michael Gross, eds., *The Custom-Made Child? Women-Centered Perspectives* (Clifton, N.J.: The Humana Press, 1981).

96. For one of the few benign views of a future world in which children are gestated artificially, see Marge Piercy, *Woman on the Edge of Time* (New York: Alfred A. Knopf, 1976).

97. See, for example, Kevles, *Eugenics,* p. 263.

Part II
Science and Social Progress

Radiation and Society

ROSALYN S. YALOW

INTRODUCTION

Radioactivity and radiation have existed from the beginning of the universe. On Planet Earth the first reactor went critical, not under the stands of a football stadium in Chicago on 2 December 1942, but about two billion years ago in a natural uranium-containing area in Gabon, Africa. However, it is less than a century since we have become familiar with ionizing radiation, with its potential in the service of humanity as well as the possible deleterious effects associated with overexposure. (Ionizing radiation is that which has enough energy to eject electrons from atoms. The term *radiation* as used here will refer to ionizing radiation.) This report reviews briefly some historical aspects of the application of radiation and radioactivity in clinical medicine and related research as well as some of what is known about the biologic effects of low-level radiation. It is obvious that there must be some selection in the studies of biologic effects to be considered here, because in a 1981 report to the Congress of the United States by the Comptroller General on the "Problems in Assessing The Cancer Risks of Low-Level Ionizing Radiation Exposure," it was stated that since 1902 the U.S. government alone has spent close to \$2 billion (approximately \$80 million per year in recent years) for investigations of the health effects of ionizing radiation and that at least eighty thousand scientific papers on the subject had been published worldwide. It was concluded in the Comptroller General's Report that while much has been learned about the carcinogenic effects of high doses of radiation exposure, uncertainty still exists as to how ionizing radiation causes cancer or how to predict the effects of exposure to low doses at low dose rates.

HISTORICAL ASPECTS

In 1895, Wilhelm Roentgen, while experimenting with cathode rays, observed fluorescence in substances placed at some distance from an opaque-paper covered Crookes tube, an apparatus used to study the conduction of electricity in gases at low pressure. Within a few weeks of intensive effort he was able to demonstrate that he had discovered "a new kind of rays," which were to be called X-rays. The most dramatic result of this experimentation, which quickly caught the lay mind, was an X-ray picture of his wife's hand recorded on a photographic film. The diagnostic possibilities of X-ray pictures were recognized at once. Almost one thousand papers and many books on the production and use of X-rays were published within the first year after the announcement of the discovery. In fact, the first X-ray journal, *Archives of the Clinical Skiagraphy,* was begun in Great Britain in 1896. (Skiagraphy is an old term, not commonly used today, for roentgenography—the taking of X-ray pictures). Through the years there have been enormous advances in X-ray diagnosis epitomized by the recent development of computerized tomography, or CAT scanning, that has revolutionized our ability to perceive structure within the human body.

In 1896 A. H. Becquerel presented to the Paris Academy of Sciences the results of his discovery of radiation emitted by uranium compounds. Following up on his work, Marie Curie, joined by her husband, Pierre, were able to purify even more highly radioactive materials from large amounts of pitchblende. In 1898 they announced the discovery of two new substances: polonium, named after her native Poland, and radium. For these discoveries Roentgen received the first Nobel prize in physics in 1901 and Becquerel and the Curies shared the 1903 prize in physics.

It was appreciated almost immediately that X-rays and the radiations from naturally radioactive substances could damage tissue. Becquerel observed a reddening of his skin under the vest pocket in which he was carrying radium. Pierre Curie deliberately exposed part of his arm to an impure radium source for ten hours and suffered a burn that took months to heal. Because radiation from X-rays and radium could destroy tissue, they were almost immediately used in the treatment of cancer. It is worth remembering that by 1920 the "war against cancer" had considerable public support and it was then thought that radium would provide the cure. Thus in 1921 when Marie Curie made her famous trip to America she was hailed as the "benefactress of the human race" for the "cancer cure" that radium was to have made possible and American women raised funds to purchase radium for her laboratory. Unfortunately it was less dramatic to recognize the long-delayed harmful effects of radiation on normal tissue. Marie Curie died of aplastic anemia in 1934 at the age of sixty-six. She was perhaps the most prestigious among the martyrs who were sacrificed because of ignorance concerning the deleterious consequences of radiation.

Carelessness with and overexposure to ionizing radiation could be attributable only in part to ignorance. Appropriate monitoring instruments were not readily available in this early period. It was only during World War II that the Manhattan Project instituted a Health Physics program for the thousands of workers involved in the development of the atom bomb. For the first time, film badges and dosimeters became generally available for monitoring purposes. At the end of the war, inexpensive radioisotopes obtained from the Oak Ridge reactor opened a new era in the use of artificially radioactive materials in clinical medicine. A new medical specialty, Nuclear Medicine, developed, which has had an enormous impact on our health and well-being. Radioimmunoassay (RIA), for the discovery of which I was awarded the Nobel prize, is a test-tube procedure employed in thousands of laboratories around the world to measure many hundreds of materials of biologic interest that are found in blood and body fluids. These substances include, among others, hormones, drugs, and proteins associated with infectious materials such as viruses and bacteria. Public health uses of RIA of considerable importance include screening of all neonates for underactivity of the thyroid gland, a disease that occurs in one in four thousand births in the United States and in one in ten births in regions of endemic goiter and which results in irreversible mental retardation if not treated before the disease is clinically recognized; and screening of blood used for transfusion for contamination with a virus that causes hepatitis, an inflammation of the liver often associated with severe morbidity and even death. Dramatic advances have also been made in methods employing radioisotopes (or radionuclides as they are now called) for imaging the human body. Not only have these methods been employed for localization of tumors, but new noninvasive dynamic techniques have been developed for studying cardiovascular disease, glucose metabolism in the brain, and receptors in the brain for drugs. Artificial radionuclides have a role in therapy as well as in diagnosis. Radioiodine (^{131}I) has been extensively used in the treatment of hyperthyroidism and thyroid cancer. Radioactive iridium (^{192}Ir) and radioiodine (^{125}I seeds encapsulated in nylon) have replaced radium and radon seeds in radiotherapy. Unfortunately, the application of radionuclides in medicine as well as the use of nuclear reactors for production of electricity are threatened by what has become an almost phobic fear of exposure to radiation at any level.

MEASUREMENT OF RADIATION EXPOSURE

Before discussing a small sampling of the tens of thousands of papers dealing with biologic effects of radiation, it is necessary to define the units used in its measurement. A rad is a unit of absorbed dose or energy absorbed per unit mass from ionizing radiation and corresponds to 100 ergs/gram, (or 0.01 joule/kg). Densely ionizing radiation such as that associated with alpha particles, protons, or fast neutrons is more effective in producing deleterious biologic effects than is the lightly ionizing radiation associated with beta-, gamma-, or X-radiation. A rem is a unit that takes into account the relative biologic effectiveness (RBE) of lightly (low linear energy transfer, LET) and densely (high LET) ionizing radiation. A rem is an absorbed dose that produces the same biologic effect as 1 rad of low LET radiation. Rad and rem are generally used interchangeably for low LET radiation. However the RBE is not a constant for any type of ionizing radiation but depends to some extent on both the radiation energy and the biologic effect under observation. For lower exposures the unit millirem is employed (1,000 mrem = 1 rem). The unit for measuring radioactivity is the Curie (Ci) and equals 37 billion disintegrations per second. The millicurie (mCi) and microcurie (μCi) are a thousandfold and a millionfold lower, respectively, and are the units more frequently used in nuclear medicine.

STANDARDS FOR RADIATION PROTECTION

By the 1920s it had become evident that there was an increased incidence of aplastic anemia, leukemia, and bone cancer among radiologists and radium workers. During that decade, national and international radiologic societies began to investigate and set standards for radiation protection. Until the early 1950s radiation protection was based on the concept of a "tolerance dose," a level below which it was believed radiation caused no harmful effects. However, at that time agencies concerned with developing protection guides and regulations utilized the concept of a linear-dose response curve, which states that a given amount of radiation produces the same number of cancers independent of the number of people who received that dose or the rate at which the dose was delivered. This would be equivalent to saying that if one hundred persons each smoke ten thousand cigarettes a year (1.5 packs per day) for twenty years, there would be the same total number of cases of smoking-induced lung cancers as if one million people smoked one cigarette a year for twenty years. Even with the current fears associated with passive smoking, no one really believes that anyone would develop lung cancer as a consequence of smoking a single pack of cigarettes during a twenty-year period. Thus it would seem that linear extrapolation hypothesis sets an upper limit, but is not a realistic formula for determination of cancer induction by low doses delivered at low dose rates. Nonetheless, in common parlance the term *person-rem* is employed. This suggests that the same absolute number of cancers would result among 100,000 people receiving 100 rem or among 100,000,000 people receiving 0.1 rem.

In 1954 the National Bureau of Standards Handbook 59, *Permissible Dose from External Sources of Ionizing Radiation,* was published. It established recommendations for radiation exposure for radiation workers and for the general public. The former were essentially to receive no more than 5 rem per year; the latter no more than 0.5 rem per year above natural background. There has been no lowering of the recommended dose levels since that time, although the general philosophy is to lower the dose as much as would be reasonably achievable.

NATURAL BACKGROUND RADIATION

It should be appreciated that environmental radiation from natural sources is the major source of radiation exposure to the inhabitants of our planet (1). Our body contains about 0.1 μCi of ^{40}K, a primordial energetic beta-emitter with a half-life of 1.3 billion years. It contributes about one-third to our natural background. Our body also contains 0.1 μCi of ^{14}C, a much weaker beta-emitter with a half-life of five thousand years, which contributes much less to our radiation exposure. Another one-third of background radiation is due to cosmic radiation and the other one-third to the natural radioactivity of soil and building materials. The average natural background whole body radiation dose in the United States is considered to be about 0.1 rem per year.

It would seem most reasonable that if we want to determine what the effects of radiation are at dose rates and doses similar to those attributable to natural background radiation, populations should be examined who live in regions of the world where background radiation is higher than that to which most of the rest of the world is exposed. Such a study was performed in China by examining 150,000 Han peasants with essentially the same genetic background and same life style (2). Half of the group lived in a region where they received an almost threefold higher radiation exposure because of radioactive soil. More than 90% of the progenitors of the more highly exposed group had lived in the same region for more than six generations. The investigation included determination of radiation level by direct dosimetry and evaluation of a number of possible radiation-related health effects including chromosomal aberrations of peripheral lymphocytes, frequencies of hereditary diseases and deformities, frequency of malignancies, growth and development of children, and status of spontaneous abortions. This study failed to find any discernible difference between the inhabitants of the two areas. The authors of this study concluded that either there may be a practical threshold for radiation effects or that any effect is so small that the cumulative radiation exposure to three times the usual natural background resulted in no measurable harm in this population after six or more successive generations. Even if the linear extrapolation were valid, one would not have expected to have detected harmful effects because the size of the exposed group was comparable to that of the Hiroshima-Nagasaki survivors and their cumulative exposures over a ten-year period were only about 10% of the acute exposure of the Japanese survivors. It should be appreciated that among the eighty-two thousand atom-bomb survivors, whose mean acute whole body dose was about 25 rem, the incidence of malignancies through 1978 was only about 6% greater than would have occurred without the radiation exposure—that is 4,500 cancer deaths would have been expected without additional radiation exposure and an additional two hundred fifty cancer deaths, ninety of which were leukemia, were estimated to be a consequence of such exposure (3). However if, as some suggest, the linear extrapolation hypothesis underestimates effects at low doses and dose rates, then the Chinese study might have discerned such effects.

In different regions of the United States there are variations in natural background radiation such that those residing in the Rocky Mountain states receive on the average approximately an additional 0.1 rem/year compared to the rest of the population. Frigerio and Stowe (4) have observed that the cancer rate in the seven states with the highest background are about 15% less than the average United States rate. A more recent study that took into account possible confounding factors such as industrialization, urbanization, and ethnicity appeared to confirm a deficit in cancer mortality in high altitude counties (5). Data such as these might suggest a protective effect of excess radiation delivered at a low dose rate, although other factors might be considered. Nonetheless, had the cancer incidence or mortality been greater in the Rocky Mountain states, radiation effects, rather than other environmental factors, would have been unequivocally declared by some to be the causative agent. In the Rocky Mountain states cumulative excess exposure averages about 1 rem for each decade of residence. Thus, even in a group as large as five million persons receiving this excess radiation exposure, genetic or life-style factors are of such overwhelming causative importance that one cannot attribute variations of cancer incidence or mortality either to advantageous or deleterious effects of low-dose/low-dose-rate radiation. There are regions of the world such as in India and Brazil where natural background radiation is up to tenfold higher than usual (~1 rem/year) and deleterious health effects have been looked for and not found (6–10). It should be appreciated that over a twenty-five-year period these exposures equal the acute exposures of the Hiroshima-Nagasaki survivors.

RADON LEVELS IN THE HOME

Soon after the discovery of radium, a mysterious emanation from the element was noticed. This proved to be a noble noninteractive gas, radon, which disintegrates with a half-life of 3.85 days followed by a series of several short-lived daughter products. By the 1930s it was appreciated that miners, in particular uranium miners, had an increased incidence of lung cancer, which was presumed to be due to elevated radon levels in the mines. It is now considered that the dose to the bronchial epithelium is largely due not to radon itself but rather to its daughter products. Until recently it was not thought that indoor levels of radon were likely to be sufficiently high to be cause for concern except perhaps in the special cases of houses built on uranium mill tailings in Colorado or on phosphate residues in Florida. However in December 1984, when a worker set off radiation alarms on his way into Pennsylvania's Limerick nuclear power plant, it was recognized that radon contamination in homes in the Reading Prong area, stretching from Pennsylvania through northwest New Jersey into New York, and in Clinton, New Jersey, and perhaps elsewhere in the country, may exceed levels found in uranium mines. As a consequence of these findings the National Council on Radiation Protection has estimated that, nationwide, 9,000 deaths annually from lung cancer might be due to radon in homes. The Federal Environmental Protection Agency had estimated the radon-related lung cancer deaths to be in the range of 6,000 to 20,000 per year, comparable to the Centers for Disease Control estimates of 5,000 to 30,000 radon-related deaths. These numbers are obtained using lung cancer risk estimates given in the BEIR III Report (11) that were based on studies of uranium miners.

According to the American Cancer Society it is estimated that there would be 89,000 male and 41,000 female lung cancer deaths in 1986 (12). Let us consider if it is likely that up to one-fourth of these lung cancer deaths could be due to indoor radon levels. Because it is unlikely that levels of radon have changed significantly with time, one can evaluate its effect from the lung cancer death rate that would obtain in a population devoid of the pulmonary insult of smoking. The age-adjusted lung cancer death rate in 1981 for males was 72 per 100,000 population; the corresponding rate for females was 21. In 1930 the male and female age-adjusted cancer death rates were 4 and 2 per 100,000 respectively (12). Because, even in 1930, some lung cancers in males were undoubtedly due to smoking during and after World War I and because there is no reason to anticipate a sex-linked difference in lung cancer, the female rate was probably closer to the true lung cancer rate for nonsmokers. Was there a marked underdiagnosis of lung cancer among women in 1930? This is not likely because the rate only slowly increased until 1960, when the effects of post–World War II smoking among women resulted in a continuous steeper rise in their lung cancer death rates. Furthermore it should be noted that the age-adjusted lung cancer incidence rate among Mormon females in Utah between 1967 and 1975 was only 4.7 per 100,000 and that of Mormon males 27 per 100,000 (13). Although the incidence for Mormon males is less than one-half that for American males during the same period, it does suggest that not all Mormons abstain from smoking. Therefore, at least some of the Mormon female lung cancer incidence in this period may be attributable to smoking and in the Mormon community underdiagnosis is unlikely since they have excellent medical care. It is therefore quite likely that the lung cancer death rate in nonsmokers should be no more than 2 to 3 per 100,000 or only about 5,000 lung deaths a year. It is extremely unlikely that all lung cancers in nonsmokers are radon-related. Therefore radon-induced lung cancer death estimates of 20,000 to 30,000 are in error by at least a factor of 10.

Evidence that the major fraction of lung cancer in nonsmokers is probably not associated with radon exposure also comes from histologic studies. The predominant histologic diagnosis of lung cancer in nonsmokers is adenocarcinoma (65%), with large-cell carcinoma accounting for about one-half the rest (14). Adenocarcinoma of the lung is rare in uranium miners; it is the small-cell undifferentiated cell types that predominate and appear to be related to cumulative radiation exposure (15). Taken together, this evidence suggests that the BEIR III estimates are too high to be used in estimates for radon-induced lung cancer in the home setting, perhaps because of the presence in uranium mines of carcinogens other than radon and its daughters or because even high LET radiation if delivered at low dose rates, as occurs in the homes with elevated radon levels, is less carcinogenic than similar radiation delivered at higher dose rates to miners. A study that might prove to be of considerable interest would be to perform autopsies on nonsmokers living in regions such as the Reading Prong to search for precancerous lesions and to determine histologic types of lung cancers and compare such findings with findings in control populations living in low radon areas.

LIMITATIONS OF CASE-CONTROL METHODOLOGY

Epidemiologic studies are of limited value in testing the validity of the linear extrapolation hypothesis for estimating effects at low doses and dose rates. Land has pointed out (16) that testing this hypothesis for radiation-induced breast cancer would require a sample size of 100 million women to be certain of an increased cancer incidence following an acute exposure of 1 rem to both breasts at age thirty-five. Because such a sample is hardly practical, a case-control approach, in which the sample consists of a fixed number of cancer cases and a fixed number of matched controls, is generally used. Land has calculated that using this cohort approach only one million women would be required to be certain of a radiation effect from 1 rem (16).

The problem using case-control approach to evaluate radiation and other potential carcinogens is that a sufficient number of subjects are never included and there is not random selection of cases and controls. Hence, the data presented often do not have statistical significance and subtle sources of bias could well account for purported observed effects. For instance, Mac-Mahon (17) reported that children born after their mothers had received one to six pelvic radiographs (average dose per radiograph was 1 rad) were 43% more likely to die of cancer in the first ten years of life than were children not irradiated in utero. Using the same case-control method of analysis Mac-Mahon et al. (18) also reported that drinking one to two cups of coffee a day introduced a relative risk of 2.6 in developing cancer of the pancreas and further suggested that coffee drinking at this level can account for more than 50% of the cases of pancreatic cancer. However, because coffee drinking is familiar and radiation is not, most people discounted his case-control analysis that appeared to prove that such modest coffee drinking is a risk factor for pancreatic cancer, particularly since the effect did not appear to be dose-related in men—the risk factor was the same, 2.6, whether consumption was one to two cups or greater than five cups a day. Perhaps because of the criticisms of his earlier paper MacMahon's group (19) subsequently conducted another case-control study and concluded that if there is any association between coffee consumption and cancer of the pancreas, it is not as strong as their earlier data had suggested (18). Nonetheless, his initial study on the association between coffee drinking and cancer of the pancreas (18) was, in a sense, less flawed than his earlier report on the association between prenatal radiation and early cancer death (17) because in the latter study there was clearly a bias in that no account was taken of the fact that the exposed mothers had medical conditions that prompted the diagnostic X-rays. A subsequent study by Oppenheim et al. (20) failed to find radiation-related effects when maternal irradiation was received on a routine basis rather than for a particular medical condition. It is not without interest that Oppenheim et al (20) reported a study in which radiation exposure due to routine pelvimetry appeared to result in a lower death rate (1.05 percent) for exposed children before the age of ten than occurred in the control group (1.42 percent).

It would seem from this sampling that flaws in case-control methodology limit the value of epidemiologic studies when possible effects are small compared to the "noise" in the system, that is, the natural variability of the end point under study.

RADIATION EXPOSURES ASSOCIATED WITH DIAGNOSTIC AND THERAPEUTIC RADIATION

Following the availability of radionuclides from Oak Ridge, ^{131}I was extensively used in the diagnosis and treatment of thyroid disease. By 1968 it was estimated that more than 200,000 hyperthyroid patients were treated with [131]I (21). Because radioiodine remains the preferred method for definitive therapy of this disease, the total number so treated has probably doubled. Thyroidal uptake of ^{131}I was the method of choice for the diagnosis of thyroid disease until RIA of thyroid-related hormones became available in the 1970s. It is estimated that 1 to 3 million people in the United States alone have received thyroidal doses in the range of 50 to 100 rems as a consequence of thyroidal uptake studies. There has been no systematic follow-up for radiation-induced malignancy in most of these several million patients. However a follow-up has been reported of a small subset of these patients. Holm et al. (22) have reported a retrospective study of over 10,000 patients in Sweden who received an average of 60 μCi[131]I and a thyroidal dose of about 60 rem between 1952 and 1965 for diagnostic purposes. Tracer studies were performed mainly on adults; only 5% of the patients were under 20 years at the time of ^{131}I administration. The expected incidence of thyroid cancer in a control population of 10,000, according to data from the Swedish Cancer Registry, was 8.3, and only 9 were observed. The mean follow-up period for the patients averaged 18 years, ranging from 10 to 25 years. The risk factors used by the United Nations Scientific Committee on the Effects of Atomic Radiation (23), which were derived from studies at high dose rates and from external X-radiation, predict an excess of 50 to 150×10^{-6} thyroid malignancies per person-rem within 25 years of radiation, with an equal number to be detected subsequently. Were this risk estimate valid, then in this group one would have expected an excess of 30 to 90 thyroid malignancies by 25 years, yet no excess was observed by 18 years. Because ^{131}I thyroidal uptake studies have generally not been performed in nonhyperthyroid patients for more than a decade, it might be of interest to extend the studies of Holm et al. (22) to a larger fraction of the several million euthyroid patients receiving ^{131}I diagnostic tests in the earlier period.

Probably the largest group ever exposed to total body radiation in the 5 to 15 rem range were ^{131}I-treated hyperthyroid patients. A study of 36,000 such patients from 26 medical centers of whom 22,000 were treated with a single dose of ^{131}I and most of the rest with surgery revealed no difference in the incidence of leukemia between the two groups (21). The average bone marrow dose was estimated to be about 10 rems, more than half of which was delivered within 1 week. The follow-up for the ^{131}I-treated group averaged 7 years, long enough to have reached the peak incidence for leukemia, as had been determined from the Hiroshima-Nagasaki experience (3). A subsequent follow-up of the hyperthyroid patients 3 years later continued to reveal no difference in leukemia rates between the two groups (24). This study emphasizes the importance of having an appropriate control group. Earlier studies had suggested that the occurrence of leukemia in hyperthyroid patients following ^{131}I therapy was 50% greater than that of the general population (25, 26). However, it appears from the study of Saenger et al. (21) that there is a 50% increased incidence of leukemia in hyperthyroidism, irrespective of the type of treatment.

It would be of interest to consider the feasibility of a large epidemiologic study of the several hundred thousand patients who have been treated with ^{131}I for hyperthyroidism. Such a study would have the potential for answering the question as to whether general whole-body radiation exposure in the 10 rem range, approximately half of which is delivered in less than 1 week, is carcinogenic. Because it appears that hyperthyroidism per se may be associated with leukemia, the appropriate control group should be, as in the study of Saenger et al. (21), patients treated with surgery. However, it may not be possible to obtain an age-matched surgically treated group because ^{131}I has become the preferred treatment. In evaluating whether hyperthyroid patients treated with antithyroid drugs until remission would be suitable as a control group, the potential of these drugs for inducing leukemia must also be considered because, in some patients, antithyroid drugs are known to produce bone marrow depression.

The question of malignancies induced by diagnostic X-radiation remains a matter of public concern. In addition to the studies of MacMahon et al. (17) and Oppenheim et al. (20) described earlier, Stewart et al. (27) reported a positive association between acute and chronic myelocytic leukemia and diagnostic X-radiation. That study was based on questionnaires given by physicians to patients with leukemia and to control subjects. The number of diagnostic procedures was then counted and a uniform quantity of radiation dose was assumed for each procedure. All information was based on subjects' recall, not on objective confirmation by medical records. The Stewart et al. study (27) should be compared with a case-control study by Linos et al. (28) of 138 cases of leukemia, representing all known cases in Olmstead County, Minnesota, between 1955 and 1974, and matched controls. This study revealed no statistically significant increase in the risk of developing leukemia after radiation doses up to 300 rads to the bone marrow when these doses were administered in small doses over long periods of time, as is the case in routine medical care. The difference between the Stewart et al. (27) study and that of Linos et al. (28) was that in the latter study the exposure to radiation administered for medical reasons had been prerecorded and was documented through careful review of records that had been entered when the radiation exposures had actually occurred, not after the diagnosis of leukemia was made. The Olmstead County experience is unique in that virtually all medical care is provided by the Mayo Clinic and one other private medical group practice and that the recordkeeping and estimations of bone marrow dose are very reliable.

Consider also other studies dealing with a possible increase in the incidence of leukemia associated with radiotherapy. Ankylosing spondylitis is a chronic and usually progressive disease of the sacroiliac joints, over 90% of the cases occurring in males in the late teens and early twenties. Radiation therapy of the affected joints had commonly been used for symptomatic relief of pain and stiffness until the studies of Court-Brown and Doll (29, 30), which reported an almost fivefold increase in leukemia in those so treated between 1935 and 1960. In these studies there appeared to be no clear relationship between the excess risk of leukemia and the estimated bone marrow radiation dose, which would appear to be a rather unusual finding. In contrast with these studies there are negative studies in which induction of leukemia as a consequence of exposure to radiation therapy was sought for and not found. Perhaps early studies (31, 32) in which no increase in leukemia in women treated for cervical cancer with either intracavitary radium, external radiation, or both were neglected because of incomplete patient follow-up. However, a report of an International Collaborative Study of more than 31,000 women with cervical cancer, of whom 90% received radiation therapy and 10% did not, revealed that 15.5 cases of leukemia were expected in the irradiated group but only 13 were observed (33). In the nonirradiated group two cases of leukemia were observed as compared with the 1.0 expected. The follow-up was long enough to have included the period of leukemia peaking observed with the Japanese atom bomb survivors (3). The consistency of these studies (31–33) would suggest that there is no detectable leukemogenic effect in patients with cervical cancer following radiotherapy. The cohort size of this study is comparable to the Court-Brown and Doll studies (29, 30). It does remain a mystery as to why radiotherapy would appear to be leukemogenic in one disease and not in another when the therapeutic doses are in the same range although not delivered to identical regions of the body.

Calculated in terms of person-rems, diagnostic X-ray examinations result in a population exposure about equal to that of natural background. However, there remains some controversy as to whether this radiation exposure is of any significance although there is general agreement that if there is any danger from diagnostic X-ray exposure, as currently employed, it is much too small to be detectable (34). It would seem obvious that if an X-ray examination is required for diagnosis of a particular medical condition, the potential benefit would exceed the risk, if any. The question as to whether mass screening programs should be employed is not that easily answered. Fifty years ago when pulmonary tuberculosis was a serious problem, mass screening examinations of the lung employing photofluorography were commonly employed. This method, which is no longer employed, was relatively inexpensive but delivered 3,000 to 5,000 mrads in air at the entrance portal. A chest X-ray study currently results in a radiation exposure 100-fold lower. However the incidence of pulmonary tuberculosis in the general population is now so low that mass screening for that disease is no longer recommended. At present the question of interest is whether screening of asymptomatic women with mammographic radiography for early detection of breast cancer is advisable. In answering this question it should be appreciated that the increased incidence of breast cancer following the acute exposure associated with Hiroshima-Nagasaki bombing was restricted to women who were under forty at the time of exposure (3) and that with equipment currently available the dose in air at the skin should be in the range of 100 and 600 mrads. Therefore it would seem that fear of radiation at this level should not be a determining factor in evaluating whether mammography is desirable for screening asymptomatic women in the age range in which cancer of the breast is most common—those over forty.

INCIDENCE OF MALIGNANCIES IN RADIATION WORKERS

It is commonly accepted that early radiation workers had an increased incidence of malignancies. However, the ignorance of potential hazards led to unbelievably careless behavior. A well-known story is that of the radium-dial workers who ingested radium by tipping the brushes with their lips. The classic report of Martland et al. (35) about the high incidence in these workers of bone sarcomas and carcinomas of the mastoids and paranasal sinuses led to an abolition of this practice. A long-term follow-up completed in 1980 of radium-dial workers revealed that among 1,260 workers who entered before 1927 there were 86 of these sarcomas and carcinomas and that among 1,794 workers who entered between 1927 and 1950 there were none (36).

What is the radiation-related incidence of malignancies in other groups of radiation workers? A report of the mortality from cancer and other causes among 1,338 British radiologists who joined radiologic societies between 1897 and 1954 revealed that in those who entered the profession before 1921, the cancer death rate was 75% higher than that of other physicians, but that those entering radiology after 1921 had cancer death rates comparable to those of other professionals (37). Although the exposures of the radiologists were not monitored, it is estimated that those who entered between 1920 and 1945 could have received accumulated whole-body doses of the order of 100 to 500 rem.

Another large group of radiation workers studies were men in the American Armed Services trained as radiology technicians during World War II and who subsequently served in that capacity for a median period of twenty-four months. Description of their training included the statement that "during the remaining two hours of this period the students occupy themselves by taking radiographs of each other in the positions taught them that day" (38). In this report it was noted that the students did not receive a skin erythema dose nor did they show a drop in white count—monitoring procedures that are insensitive to acute doses less than 100 rem. Although the cumulative exposures of these radiology technicians were not monitored, the radiation exposures of technologists at a more modern installation, Cleveland Clinic, were monitored in 1953 and found to be in the range of 5 to 15 rem/year (39). Army technologists a decade earlier probably received as much as 50 rem or more during their training and several years of service. Yet, a twenty-nine-year follow-up of these six thousand five hundred radiology technicians revealed no increase in malignancies when compared with a control group of similar size consisting of Army medical, laboratory, or pharmacy technicians (40, 41).

There have been well-publicized studies suggesting that linear extrapolation greatly underestimates carcinogenesis in nuclear reactor-related radiation exposure. Let us consider the facts in two of these cases, one at the Portsmouth Naval Shipyard and another involving Hanford radiation workers. In 1977 Dr. Thomas Najarian, together with the *Boston Globe,* investigated the rumors that cancer was rampant among nuclear workers at the shipyard. The *Globe* story appeared before the work was submitted for publication and was sufficiently sensational to have resulted in Congressional hearings. The preliminary paper published by Najarian and Colton (42) was based on interviews, with the help of five Globe reporters, with next of kin of workers who were reported to have been "exposed" or "unexposed." They claimed to have found 6 leukemia deaths among 146 "exposed" worker deaths whereas only 1.1 would have been expected. An extensive reinvestigation by the National Institute of Occupational Safety on Health (NIOSH), in which workers were classified as "exposed" or "unexposed" on the basis of radiation exposure records, revealed that among the 4,566 deaths occurring between 1952 and 1977, there was no increase in either leukemia or cancer among the exposed workers (43). There was no trend in increasing cancer deaths with dose up to a maximum of 90 rem. This study had the power to detect a doubling of the leukemia rate.

Another report that has received publicity far beyond its scientific merit is the Mancuso study of workers at the Hanford Laboratories, the site of several reactors of the Atomic Energy Commission (AEC). The history of this study dates back to 1964 when Dr. T. Mancuso was awarded a contract to investigate for the AEC the health of these workers. Dr. Sanders, a statistician, and Dr. Brodsky, a health physicist, were co-investigators in this project. Annual project reports for many years suggested only that there were negative findings regarding a link between cancer and radiation, and Sanders and Brodsky left the project. There were no papers published during the period in which they were involved in the analysis of the data. In about 1976, Mancuso was joined by the pediatrician Dr. A. Stewart and the statistician G. Kneale who had acquired a reputation for their studies on relationships between diagnostic X-rays and childhood cancers. Together they wrote a controversial paper purporting to show that workers at Hanford had a statistically significant increase in the incidence of two types of cancer, multiple myeloma and cancer of the pancreas (44). In their paper they reported that the mean cumulative radiation dose for Hanford workers who subsequently died from cardiovascular disease was 1.05 rads; for solid tumors, 1.3 rads; for leukemias and lymphomas, 2.2 rads. This excess radiation exposure is quite comparable to the excess received by living in Colorado for ten to twenty years—and Colorado has a low cancer death rate. The evidence in the Mancuso report that has not been widely publicized was that Hanford workers receiving the highest radiation doses (greater than 15 rem) had a lower death rate from all causes and from all malignant neoplasms than expected in a control population. However, because of the small number of workers who received this exposure and the small number of cancer deaths in this group (a total of 14 cancer deaths compared to 24 expected), the distribution among the different malignancies appeared to have a pattern not identical with that found in much larger groups. Subsequently, both Brodsky and Sanders, who initially collaborated with Mancuso, have been highly critical of the Mancuso paper. An independent analysis by Gilbert and Marks is most revealing (45). The positive correlation purported to have been demonstrated in the Mancuso report appears to be due to 3 deaths from pancreatic cancer in workers receiving more than 15 rem cumulative exposure. However, according to Gilbert and Marks (45) this diagnosis had been confirmed only in 1 case. Furthermore, it should also be noted that in the survivors of the Hiroshima-Nagasaki bombings there was no positive link between pancreatic neoplasms and radiation (3). The second category of excess cancer deaths was reported to be multiple myeloma, which included 3 cases compared with an expected number of 0.6 (40). Whether this excess of 2 deaths in this category represents a statistical variation or the effect of another carcinogen cannot be determined. Nonetheless, because among those receiving a cumulative exposure of 15 rems the observed number of subjects with malignancies was only 14 compared with an expected 24 in a control population, one could be tempted to conclude that radiation at this level is protective against malignancies.

POTENTIAL FOR INCREASE IN MALIGNANCIES ASSOCIATED WITH TESTING OF NUCLEAR WEAPONS

There has been considerable publicity given to problems of the so-called Atomic Veterans. Caldwell et al. (46) reported an increased incidence of leukemia among 3,200 men who participated in Operation Smoky, a nuclear explosion at the Nevada Test Site in 1957. Stimulated by this report the Medical Follow-up Agency of the National Research Council studied the mortality and causes of death of a cohort of 46,186 participants, about one-fifth of the total number of participants in one or more of five atmospheric nuclear tests (47). The reanalysis confirmed that among the 3,500 participants at Operation Smoky the standardized mortality ratio (SMR) for leukemia was 2.5—that is, there were 10 observed leukemia deaths and only 3.97 expected. Only 1 of those 10 had received an exposure in excess of 3 rem. For all other cancers the SMR's were less than 1.0. It is of particular interest that among participants in Operation Greenhouse at a Pacific Test Site in 1951, with a cohort size of almost 3,000, the expected leukemia mortality was 4.43 yet only 1 was observed for an SMR of 0.23. For the other malignancies, where the numbers involved are much larger, the SMRs are in the range of 0.7 to 0.9. If we examine the entire cohort of 46,000 we find the SMR for all malignancies is 0.84 and for leukemia it is 0.99. The excess SMR for leukemia at Operation Smoky and the equivalently decreased SMR at Operation Greenhouse are typical aberrations attributable to small-number statistics. Could one have expected an increase in leukemia at Operation Smoky? Because only 1 of the veterans with leukemia was reported to have received more than 3 rem, the probability of observing a true increase in leukemia would require a gross underestimate of the radiation dose received by the participants. However a committee, chaired by Dr. Merril Eisenbud for the National Research Council, reviewed the methods used to assign radiation doses to service personnel at nuclear tests and concluded that the methods were reasonably sound but that doses assigned to the test participants were probably somewhat higher, not lower, than the actual doses received (48). This report also reviewed a number of studies that estimated radiation exposure from internally deposited radionuclides and concluded that these did not add significantly to the external exposure.

There have also been reports of increases in malignancies among civilians exposed to fallout from nuclear testing. In 1979 Lyon et al. (49) reported that leukemia mortality in children was increased in those counties in Utah receiving high levels of fallout from the atmospheric nuclear testing conducted between 1951 and 1958 compared with the mortality in low fallout counties and in the rest of the United States. In the 1944–50 and 1959–75 periods, the leukemia mortality in the so-called high fallout regions was considerably less than in the rest of Utah and the United States. In addition, if one considers the sum of childhood malignancies (leukemia plus other cancer deaths), there appears to be generally downward trend, from 1944 to 1975, with the drop in the high-fallout counties being somewhat greater than in the low-fallout counties, although if the standard deviations had been included the differences would not have been significant. The news headlines following interviews with Dr. Lyon would have been less sensational if he had stated that his data had shown no relation between the totality of childhood malignancies and the atomic tests of the 1950s, rather than selectively reporting an inconclusive study of the relation between leukemia and fallout. Lyon's paper (49) was criticized in the same and later issues of the same journal by several biostatisticians (50–52). In general, their criticisms were related to the apparent underreporting or misdiagnosis in the earlier cohort and to errors in small sample analysis. For instance, Bader (51) presented a year-by-year listing of leukemia cases in Seattle–King County, which has a larger population than the southern Utah counties, and noted that there were only 2 cases in 1959 and 20 in 1963 among the 217 cases reported from 1950 to 1972. Thus, a tenfold difference in annual incidence rates when the number of cases is small simply represents statistical variation. Although the yearly distribution of leukemia cases has never been reported in any of Lyon's papers or in the associated publicity, it was tabulated in the Comptroller General's Report previously referred to. It is of interest that the so-called excess of leukemia cases reported by Lyon et al. (49) was due to a clustering of 13 cases in 1959 and 1960. In fact, 22 of the 32 leukemia cases occurred between 1951 and 1960—that is, during the first 10 years of testing. Because there is a several-year latency period between radiation exposure and induction of leukemia, if the excess leukemia deaths were a consequence of nuclear testing in the 1950s, they are more likely to have occurred after 1960 rather than before. Furthermore, a new estimation of external radiation exposure of the Utah population based on residual levels of ^{137}Cs in the soils has shown that the mean individual exposure in what Lyon deemed to be the "high fall-out counties" was 0.86 ± 0.14 rad compared to 1.3 ± 0.3 rad in the "low fall-out counties" (53). Even in Washington County, the region in which the fallout arrived the earliest (less than 5 hours after the test) the estimated exposure to its 10,000 population averaged only 3.5 ± 0.7 rads—quite comparable to natural background radiation in that region over a 20-year period. Thus on the basis of the Japanese experience the exposure from fallout was too low to expect an increase in leukemia and a careful perusal of Lyon's data would suggest that none was found.

Measurement of residual radioactivity in the soil permits reconstruction of the external gamma-ray exposure of the population downwind from the nuclear test sites. However there has also been concern with inhaled or ingested radionuclides. Therefore a ten-year program was initiated in 1970 to determine levels of radionuclides in adults and children from families residing in communities and ranches surrounding the Nevada test site (54). Monitoring was done by whole body counting, which would determine ^{60}Co, ^{131}I, and ^{137}Cs and by testing of urine for ^{239}Pu and ^{238}Pu. ^{131}I was included in the monitoring because the underground Baneberry test of 1970 was vented and there was some concern about ^{131}I contamination of the milk supply. Monitoring revealed that the estimated thyroidal dose to members of a family using the contaminated milk was about 6 mrem, about 1/10,000 of the thyroidal dose routinely received by the several million people who had received diagnostic doses of ^{131}I for testing of thyroid function between 1948 and 1968. ^{137}Cs was the major fission product found in the off-site population and its concentration was similar to that found in people living elsewhere in the United States. No clear difference was found between persons living in the windward and leeward sides of the Nevada Test Site (54).

In view of the failure to demonstrate significant external or internal radiation doses resulting from fallout from nuclear testing, what then could account for the remarkably high incidence of many types of cancer downwind from the Nevada test site reported by Johnson (55)? Johnson reports from a series of personal or oral communications that doses received by residents were in the range of hundreds to thousands of rads. He does not reference or discuss the earlier published scientific reports demonstrating the failure to detect significant contamination of the people or soil in the fallout areas (53, 54). However, he does reference a "News and Comment" article in *Science* (56) that reports on a lawsuit against the government of the United States in which allegations were made that the government was grossly underestimating human radiation exposure following nuclear bomb testing. Even if there were some underestimation of such exposure, it is most remarkable that through 1975 he reports there were 288 cancer deaths in the so-called exposed group (4,125 people) compared with 179 expected (55) whereas the Japanese experience was only an extra 250 cancer deaths over the 4,500 expected by 1979 in a population of 82,000 (3). The Johnson report (55) is certainly inexplicable.

USEFULNESS OF ANIMAL STUDIES IN PREDICTION OF HUMAN RADIATION-INDUCED CARCINOGENESIS

Animal studies have certain advantages: the animals are inbred and are not subject to the genetic and environmental variability of a human population; at present it is possible to expose animals but no humans to graded radiation doses at different dose rates. The inherent limitation of such studies is that it would be enormously expensive to maintain the large groups of animals that would be required to evaluate effects at truly low doses and dose rates. The conclusion of many studies of different tumors in different animals is that for a given total dose there was generally decreased tumorigenesis when the radiation was delivered at a lower dose rate, but that the reduction factor was dependent both on the tumor type and the species of animals (57). None of these studies have been performed at truly low dose rates. For instance, the studies by Ullrich and Storer (58) on tumorigenesis in mice revealed that when ^{137}Cs gamma-ray irradiation is delivered at 8.3 rad/day (i.e., 25,000 times natural background), there is a threshold of about 50 rads before an increased incidence of ovarian tumors or thymic lymphomas is observed. The threshold appeared to be no more than a few rads when the irradiation dose rate was 45 rad/min. Studies at very low dose rates—for instance, at about 100 times natural background—would require an enormous number of animals and are not practical.

CONCLUSION

From the studies described in this review it is evident that epidemiologic studies cannot produce meaningful data about the existence of a threshold for radiation effects. Molecular and cellular studies may or may not give some insight about molecular or cellular effects but cannot answer important questions about repair mechanisms in the intact animal or human when radiation is delivered at low dose rates. At present there are no really good ideas that would permit a breakthrough in the field of evaluation of potential effects of low-level radiation delivered at a low dose rate.

The disagreement in the low-level radiation field is about hypothesis, not about observable facts. One could not determine the validity of Newton's Laws at subatomic dimensions until the tools became available. However, in that case there was no need to make policy decisions based on extrapolation. In the case of low-level radiation effects, public policy decisions need to be made in the absence of scientific evidence. It should be appreciated that these are arbitrary decisions based on philosophy, not fact, and may well change because of political or other considerations.

In conclusion a quote from the National Council (NCRP) Report no. 43 on Radiation Protection Philosophy is relevant. "The indications of a significant dose rate influence on radiation effects would make completely inappropriate the current practice of summing of doses at all levels of dose and dose rate in the form of total person-rem for purposes of calculating risks to the population on the basis of extrapolation of risk estimates derived from data at high doses and dose rates. . . . The NCRP wishes to caution governmental policy-making agencies of the unreasonableness of interpreting or assuming 'upper limit' estimates of carcinogenic risks at low radiation levels, derived by linear extrapolation from data obtained at high doses and dose rates, as actual risks, and of basing unduly restrictive policies on such an interpretation or assumption. Undue concern, as well as carelessness with regard to radiation hazards, is considered detrimental to the public interest."

REFERENCES

This study was supported in part by the Medical Research Program of the Veterans Administration.

1. "Natural background radiation in the United States." *NCRP Report,* no. 45. National Council of Radiation Protection, Washington, D.C. (1975).

2. High Background Radiation Research Group, China. "Health survey in high background radiation areas in China." *Science* 209:877–80 (1980).

3. Kato, H., Schull, J. "Cancer mortality among atomic bomb survivors 1950–78." *Radiat. Res.* 90: 395–432 (1982).

4. Frigerio, N. A., and Stowe, R. S. "Carcinogenic and genetic hazard from background radiation." In: *Biological and Environmental Effects of Low-Level Radiation,* Vienna, International Atomic Energy Agency (1976), pp. 385–93.

5. Amsel, J., Waterbor, J. W., Oler, J., Rosenwaike, I., and Marshall, K., "Relationship of site-specific cancer mortality rates to altitude." *Carcinogenesis* 3:461–65 (1982).

6. Barcinski, M. A., Abreu, M.-D. A., DeAlmeida, J. C. C., Naya, J. M., Fonseca, L. G., and Castro, L. E., "Cytogenetic investigation in a Brazilian population living in an area of high natural radioactivity." *Am. J. Hum. Genet.* 27:802–6 (1975).

7. Cullen, T. L., "Dosimetric and cytogenetic studies in Brazilian areas of high natural activity." *Health Phys.* 19:165–66 (1970).

8. Gruneberg, H., Bains, G. S., Berry, R. J., Riles, L., Smith, C. A. B., and Weiss, R. A., "A search for genetic effects of high natural radioactivity in South India." *Med. Res. Council Special Report Series,* no. 307, H.M.S.O. London, (1966).

9. Gopal-Ayengar, A. R., Sundaram, K., Mistry, K. B., Sunta, C. M., Nambi, K. S. V., Kathuria, S. P., Basu, A. S., David, M., "Evaluation of the long-term effects of high background radiation on selected population groups on the Kerala Coast." *Peaceful Uses of Atomic Energy* (1972), pp. II: 31–51.

10. Ahuja, Y. R., Sharma, A., Nampoothiri, K. U. K., Ahuja, M. R., and Dempster, E. R., "Evaluation of effects of high natural background radiation on some genetic traits in the inhabitants of monazite belt in Kerala, India." *Hum. Biol.* 45: 167–79 (1973).

11. BEIR III Report, *The Effects on Populations of Exposure to Low Levels of Ionizing Radiations: Committee on the Biological Effects of Ionizing Radiations, Assembly of Life Sciences.* National Academy of Sciences, Washington, D.C., (1980).

12. *Ca–A Cancer Journal for Clinicians* (Published by the American Cancer Society) 36: 14–17 (1986).

13. Lyon, J. L., Gardner, J. W., and West, D. W., "Cancer incidence in Mormons and non-Mormons in Utah during 1967–75." *J. Natl. Cancer Inst.* 65: 1055–61 (1980).

14. Garfinkel, L., Auerbach, O., and Jouber, L., "Involuntary smoking and lung cancer: a case-control study." *J. Natl. Cancer Inst.* 75: 463–69 (1985).

15. Soccomanno, G., Archer, V. E., Auerbach, O., Kuschner, M., Saunders, R. P., and Klein, M. G., "Histologic types of lung cancer among uranium miners." *Cancer* 71: 158–523 (1971).

16. Land, C. E., "Estimating cancer risks from low doses of ionizing radiation." *Science* 209: 1197–203 (1980).

17. MacMahon, B., "Prenatal x-ray exposure and childhood cancer." *J. Natl Cancer Inst.* 28: 1133–91 (1962).

18. MacMahon, B., Yen, S., Trichopoulos, D., Warren, K., and Nardi, G., "Coffee and cancer of the pancreas," *N. Engl. J. Med.* 304: 630–33 (1981).

19. Hsieh, C.-C., MacMahon, B., Yen, S., Trichopoulos, D., Warren, K., Nardi, G., "Coffee and pancreatic cancer (Chap. 2)." *N. Engl. J. Med.* 315:587–89 (1986).

20. Oppenheim, B. E., Griem, M. L., Meier, P., "The effects of diagnostic x-ray exposure on the human fetus: an examination of the evidence." *Diagn. Radiol.* 114: 529–34 (1975).

21. Saenger, E. L. Thoma, G. E., and Tompkins, E. A., "Incidence of leukemia following treatment of hyperthyroidism." *J.A.M.A.* 205: 855–62 (1968).

22. Holm, L.-E, Lundell, G., and Wallnder, G., "Incidence of malignant thyroid tumors in humans after exposure to diagnostic doses of Iodine-131: 1. Retrospective cohort study." *J. Nat. Cancer Inst.* 64: 1055–1059 (1980).

23. United Nations Scientific Committee on the Effects of Atomic Radiation, *Sources and Effects of Ionizing Radiation.* New York, 1977, pp. 3671–432.

24. Saenger, E. L., Tompkins, E., and Thoma, E. G., "Radiation and leukemia rates." *Science* 1711096–98 (1971).

25. Pochin, E. E., "Leukemia following radioiodine treatment of thyrotoxicosis." *Br. Med. J.* 2: 1545–48 (1960).

26. Werner, S. C., Gittleshon, A. M. and Brill, A. B., "Leukemia following radioiodine therapy of hyperthyroidism." *J.A.M.A.* 177: 646–48 (1961).

27. Stewart, A., Pennybacker, W., and Barber, R., "Adult leukemias and diagnostic x-rays." *Br. Med. J.* 2:822–90 (1962).

28. Linos, A., Gray, J. E., Orvis, A. L., Kyle, R. A., O'Fallon, W. M., and Kurland, L. T., "Low dose radiation and leukemia." *N. Engl. J. Med.* 302: 1101–5 (1980).

29. Court-Brown, W. M., and Doll, R., "Leukemia and aplastic anemia in patients irradiated for ankylosing spondylitis." *Med. Res. Counc. Spec. Rep. (London)* 295: 1–135 (1957).

30. Court-Brown, W. M., "Mortality from cancer and other causes after radiotherapy for ankylosing spondylitis." *Br. Med. J.* 2: 1327–32 (1965).

31. Hutchinson, G. B., "Leukemia in patients with cancer of the cervix uteri treated with radiation: a report covering the first 5 years of an international study." *J. Natl. Cancer Inst.* 40:951–82 (1968).

32. Simon, N., Bruger, M., Hayes, R., "Radiation and leukemia in carcinoma of the cervix." *Radiology* 74: 905–11 (1975).

33. Boice, J. D., and Hutchinson, G. B., "Leukemia in women following radiotherapy for cervical cancer: ten-year follow-up of an international study." *J. Natl. Cancer Inst.* 65: 115–29 (1980).

34. Boice, J. D., "The danger of X-rays-real or apparent?" *N. Engl. J. Med.* 315: 828–30 (1986).

35. Martland, H. S., Conlon, P., and Knef, J. P., "Some unrecognized dangers in the use and handling of radioactive substances." *J.A.M.A.* 85: 1769–76 (1925).

36. Rowland, R. E., and Lucas, H. F., Jr., "Radium-dial workers." In: *Radiation Carcinogenesis: Epidemiology and Biological Significance,* ed. Boice, J. D., Jr., and Fraumeni, J. F., Jr., Raven Press, New York (1984), pp. 231–40.

37. Smith, P. G., and Doll, R., "Mortality from cancer and all causes among British radiologists," *Br. J. Radiol.* 54: 187–94 (1981).

38. McCaw, W. W., "Training of x-ray technicians at the School for Medical Department Enlisted Technicians." *Radiology* 42: 384–88 (1944).

39. Geist, R. M., Jr., Glasser, O., and Hughes, C. R., "Radiation exposure survey of personnel at the Cleveland Clinic Foundation." *Radiology* 60: 186–91 (1953).

40. Miller, R. W., and Jablon, S., "A search for late radiation effects among men who served as x-ray technologists in the U.S. Army during World War II." *Radiology* 96: 269–74 (1970).

41. Jablon, S., and Miller, R. W., "Army technologists: 29-year follow-up for cause of death." *Radiology* 126: 677–79 (1978).

42. Najarian, T. and Colton, T., "Mortality from leukemia and cancer in shipyard nuclear workers." *Lancet* 1: 1018–20 (1981).

43. Rinsky, R. A., Zumwalde, R. D., Waxweiler, R. J., et al., "Epidemiologic study of civilian employees at the Portsmouth Naval Shipyard." *Lancet* 1: 231–35 (1981).

44. Mancuso, T. F., Stewart, A., and Kneale, G., "Radiation exposure of Hanford workers dying from cancer and other causes." *Health Phys.* 33: 369–85 (1977).

45. Gilbert, E. S. and Marks, S., "An analysis of the mortality of workers in a nuclear facility." *Radiat. Res.* 79: 122–48 (1979).

46. Caldwell, G. G., Kelly, D. B., and Heath, C. W., Jr., "Leukemia among participants in military maneuvers at a nuclear bomb test." *J.A.M.A.* 244: 1575–78 (1980).

47. Robinette, C. D., Jablon, S., Preston, T. L., *Studies of Participants in Nuclear Tests: Report to the National Research Council. Mortality of Nuclear Weapons Test Participants.* Natl. Acad. Press, Washington, D.C. (1985).

48. *Review of the Methods Used to Assign Radiation Doses to Service Personnel at Nuclear Weapons Tests.* Board on Radiation Effects Research, Commission on Life Sciences, Natl. Acad. Press, Washington, D.C. (1985).

49. Lyon, J. L., Klauber, M. R., Gardner, J. W., and Udall, K. S., "Childhood leukemias associated with fallout from nuclear testing." *N. Engl. J. Med.* 300: 397–402 (1979).

50. Land, C. E., "The hazards of fallout or of epidemiologic research?" *N. Engl. J. Med.* 300: 431–32 (1979).

51. Bader, M., "Leukemia from atomic fallout." *N. Engl. J. Med.* 300: 1491 (1979).

52. Enstrom, J. E., "Editorial." *N. Engl. J. Med.* 300: 1491 (1979).

53. Beck, H. L., and Krey, P. W., *"External Radiation Exposure of the Population of Utah from Nevada Weapons Tests.* DOE/EML 401 (DE8201042), National Technical Information Service, U.S. Department of Energy, NY, (1982), p. 19.

54. Patzer, R. G., and Kaye, M. E., "Results of a surveillance program for persons living around the Nevada test site—1971 to 1980." *Health Physics* 43: 791–801 (1982).

55. Johnson, C. J., "Cancer incidence in an area of radioactive fallout downwind from the Nevada test site." *J.A.M.A.* 251: 230–36 (1984).

56. Smith, R. J., "Atom bomb test leave infamous legacy." *Science* 218: 266–69 (1982).

57. *Tumorigenesis in Experimental Laboratory Animals.* NCRP Report, no. 64. National Council on Radiation Protection and Measurements. Washington, D.C. (1980).

58. Ullrich, R. L., and Storer, J. B. "Influence of dose, dose rate and radiation quality on radiation carcinogenesis and life shortening in RFM and BALB/c mice," In: *Late Biological Effects of Ionizing Radiation,* Vol. II, IAEA/STI/PUB/489, International Atomic Energy Agency, Vienna, p. 95 (1978).

In the Clutches of Daedalus: Science, Society, and Progress

SAL RESTIVO

INTRODUCTION

My objective in this paper is to articulate an agenda that has been more or less implicit in my earlier work on the sociology of objectivity—that is, to explore the affinities between anarchism and progressive inquiry. My discussion of the relationship between modern science and social progress will serve as a vehicle for introducing the conjecture that wedding anarchy and inquiry is essential for facilitating the twin processes of social and epistemic progress. The term *anarchy* as I use it is a radical extension of the basic ideas of democracy and especially participatory democracy. Anarchy in this sense is distinguished from democracy and all other egalitarian and liberatory social forms by an uncompromising opposition to the principle of Authority, and to all constraints rooted in institutions organized in terms of that principle.

In 1923, the biochemist J. B. S. Haldane published an essay titled *Daedalus, or Science and the Future*. Haldane painted a picture of an attractive future society created by applying science to the promotion of human happiness.[1] Bertrand Russell replied to Haldane in an essay on *Icarus, or the Future of Science*. Russell wrote that much as he would like to agree with Haldane's forecast, his experience with statesmen and governments forced him to predict that science would be used "to promote the power of dominant groups rather than to make men happy." Daedalus taught his son Icarus to fly, but warned him not to stray too close to the sun. Icarus ignored the warning and plunged to his death. Russell warned that a similar fate would "overtake the populations whom modern men of science have taught to fly."[2] Russell is more pessimistic than Haldane, but he is hesitant about holding Daedalus—and modern men of science—responsible for the fates of the individuals and societies they instruct. Who *was* Daedalus, and *what* is modern science?

According to Greek mythology, Daedalus was the first mortal inventor. His career is marked by fantastic ingenuity coupled with jealousy, intrigue, and murder. He invents a machine that allows Minos's queen Pasiphae to copulate with a bull. The issue of this affair is hidden away in a labyrinth designed by Daedalus, and later killed by Theseus who uses a device invented by Daedalus to negotiate the labyrinth; Theseus also steals Pasiphae's daughter in the bargain. Daedalus eventually falls out of favor with the king and is imprisoned in a tower. Eventually, Daedalus fashions wings out of feathers and wax for himself and his son Icarus. Their escape from the tower is marred when Icarus ignores his father's admonitions, strays too close to the sun, and plunges to his death when the intense heat melts the wax on his wings. Daedalus survives, bitter and lamenting his own genius. Daedalus has a nephew, Talos, whose talent as an inventor he envies. Among Talos's inventions is the first saw. But Daedalus claims *he* invented the first saw. He resolves this earliest priority dispute (and the problem of having to contend with a rival) by pushing Talos off a tower. *This* Daedalus is, in fact, an appropriate symbol for modern science. Modern science, coupled with modern technology, has helped to fashion and sustain modern industrial, technological society.

"Modern science" is a social institution organized around the social role of the scientist. It took root in the European Scientific Revolution of the sixteenth and seventeenth centuries, and crystallized as a social institution in the nineteenth century. It has undergone transformations in the scale of its activities and its power relative to other social institutions. These transformations have been coincident with professionalization and bureaucratization internally (affecting the social structure of scientific activity), and development and change in the structure and power of the state externally. The terms *internal* and *external* are used here in a sociological sense, and refer to the degree to which scientific activities are organizationally and institutionally autonomous relative to other social activities. The social role of the scientist and the power and prestige of the physical sciences are phenomena of modern—and particularly twentieth-century—science. Modern science is therefore best understood as an *organizational* phenomenon, and not an intellectual one.[3]

Scientific inquiry was not invented in the Scientific Revolution. The Scientific Revolution organized the human or cultural capacity for inquiry and for turning the results of inquiry into material results and solutions to problems. In this process, that capacity was organized and focused in ways that led to a strategy for inquiry that stressed "laws" rather than "necessities" of nature, valued what could be quantified over what had to be dealt with qualitatively, and approached objects of inquiry (including "nature") as things to be dominated and exploited rather than as aspects of an ecological whole that we needed to understand to facilitate human survival and cultural development. All of this was consistent with and paralleled the social, economic, political, and religious changes that we know as the Industrial Revolution. The result was an alienating and alienated mode of inquiry. Modern science emerged as "the mental framework of a world defined by capital accumulation" (Morris Berman), and the mode of cognition of industrial society (Ernest Gellner).[4] Industrial processes and products became a dominant part of the cultural landscape, and played an increasingly prominent role in fashioning our everyday lives. They became the "chief factor" in shaping our science; we have learned to think, Thorstein Veblen observed, "in the terms in which the technological processes act."[5] Science, scientists, and images and symbols of science are all *commodities* in the modern world.

I want to describe the features of modern science in more detail in order to establish the point that modern science *as a social institution,* while it has historically been intimately linked with the "idea of progress," has done little to facilitate *social progress.* Social progress is "measured" by the extent to which human beings lead lives, individually and collectively, characterized by freedom from want and authority, and by opportunities to develop and exercise their creative and critical talents and abilities.

MODERN SCIENCE AND SOCIETY

The social problems of modern science are masked by icons, myths, and ideologies. Icons—Archimedes drawing pretty figures in the dust, Newton searching for shapely pebbles at the beach, and Einstein riding light beams in his mind—are objects of uncritical devotion. The myth of pure science is a cornerstone of modern science as a house of worship. And the ideologues of modern science have persuaded many of us to demarcate "science" and "technology," and to blame the latter for our social and environmental ills. No wonder, then, that it is difficult for us to see the alienative aspects of scientific work, and the connections between modern science and technology, ruling elites, state interests, and God as the symbol of a moral order. What realities lie behind these icons, myths, and ideologies? To begin to answer this question, we can turn first to some observations by an ambivalent sociologist of science, Robert Merton. For it was Robert Merton, the prime mover in functionalist sociology of science, and an ideologue of modern science, who early in his career defended Boris Hessen's Marxist conjectures on the social and economic roots of Newton's *Principia* against G. N. Clark's effort to preserve at least some of the purity of Newtonian science.[6]

Merton pointed out the importance of distinguishing "the personal attitudes of individual men of science from the social role played by their research." A variety of motives, Merton argued, is compatible with "the demonstrable fact that the thematics of science in seventeenth century England were in large part determined by the social structure of the time." He also opposed Sombart's contention that science and technology were almost completely divorced in the seventeenth century. On the specific question of

Newton's motives and the social relations of science in seventeenth-century England, Merton writes:

> Newton's own motives do not alter the fact that astronomical observations, of which he made considerable use, were a product of Flamsteeds's work in the Greenwich Observatory, which was constructed at the command of Charles II for the benefit of the Royal Navy. Nor do they negate the striking influence upon Newton's work of such practically-oriented scientists as Halley, Hooke, Wren, Huyghens and Boyle. Even in regard to motivation, Clark's thesis [regarding the primacy of disinterestedness among English scientists of this period] is debatable in view of the explicit awareness of many scientists in seventeenth century England concerning the practical implications of their research in pure science. It is neither an idle nor unguarded generalization that *every English scientist of this time* who was of sufficient distinction to merit mention in general histories of science at one point or another explicitly related at least some of his scientific research to immediate practical problems.[7]

To appreciate the significance of Merton's argument regarding the social relations of science and the motives of individual scientists, it is useful to examine the idea of "pure science." An understanding of the sociology of pure science is a prerequisite for any critical study of the relationship between science and social progress. Let me begin by considering the two most important icons from before and after the period of the Scientific Revolution during which Newton lived and worked, and for which he is the preeminent iconographic representation. Newton's counterparts from the ancient and the contemporary world are Archimedes and Einstein.

Archimedes, like Newton, played down his role in practical affairs; at least that is the report we get from Plutarch. According to Plutarch, Archimedes "placed his whole ambition in those speculations in whose beauty and subtlety there is no admixture of the common needs of life." Plutarch, of course, was writing nearly three hundred years after Archimedes' death. The distinguished historian of Greek mathematics T. C. Heath supports Plutarch's view by noting that Archimedes wrote only one "mechanical" treatise, the lost work *On Sphere-Making*. Heath equates "mechanical" with "construction," for the lost manuscript on sphere-making deals with the construction of a sphere representing motions of bodies in the heavenly system. Construction, however, is not the only sort of mechanical interest that can be opposed to "pure contemplation." In other works (some lost), Archimedes deals with such practical matters as the calendar, optics, centers of gravity, balances, and levers. The fact that he was a great inventor, and that some of his inventions were designed for political and military purposes, cannot (Plutarch, Heath, and Archimedes himself notwithstanding) be ignored as "incidental." It is unreasonable to suppose that Archimedes could completely detach his mechanical interests and talents from his interests and talents in so-called pure mathematics. The coexistence of these talents and interests, amply documented, is sufficient grounds for arguing that Archimedes' mathematics was not a product of "pure contemplation."[8]

In his book on *Method,* discovered in 1906, Archimedes outlines the mechanical bases of his formal ("pure") geometric proofs. His tendency to suppress the "vulgar" roots of the results he presented in a logical format for public consumption is not an unusual strategy in the history of mathematics. Biography offers some clues regarding the roots of this strategy. Archimedes was the son of an astronomer (Pheidias), and an intimate (perhaps even a relative of) King Hieron. His *achieved* social position, at least, and the fact that he was in a position to generalize the generalizations of earlier mathematical workers, and then generalize his own generalizations (thus producing relatively high levels of abstraction), could easily have led him and admiring biographers (ulterior motives aside) to emphasize that his inquiries were not prompted by "vulgar" considerations. And whatever Archimedes' motives in any particular situation, they cannot alter the fact of his relationship to the political and military authorities of his city, a relationship that tells us something about the ties between science and society.

The case of Albert Einstein, the most prominent icon in twentieth-century science, is more complicated than the cases of Archimedes and Newton with respect to the relationship between individual motives and social roles. Einstein worked in an era of professionalized science. Twentieth-century science is more highly professionalized and bureaucratized than earlier forms of science or inquiry. It is thus easier for individual scientists to work in apparent dissocation from the practical concerns of everyday life and vulgar political and economic interests because they are shielded by complex institutional relationships. It is therefore crucial to examine the "scientific community's" relationship to the wider society, and to the state, in order to understand the social role of an individual scientist.

Einstein's activities illustrate how the scientific community—through its own internal social structure and its ties to state interests—can protect and provide for its members, and even provide niches within which one can engage in the sort of private thinking sometimes labeled "pure contemplation." Einstein wrote:

> I believe with Schopenhauer that one of the strongest motives that leads men to art and science is escape from everyday life with its crudity and hopeless dreariness, from the fetters of one's own ever-shifting desires.[9]

But Einstein's social role—and more generally the relationship between science and society—is revealed in the public relations of Albert Einstein. There is no need to impugn Einstein's motives or his humanitarian spirit to recognize that there is something sinister in all of those photographs showing Einstein posing with kings, queens, prime ministers, and presidents. The shadow of Adolf Hitler that darkens these photographs should not lead us to make the mistake of viewing the states represented in them as benevolent; it should not keep us from seeing that what is sinister about these photographs is not what they tell us about Einstein and King Albert or Einstein and President Harding, but rather what they tell us about science and the state. Let me pursue this further by turning to a brief exploration of the more general relationships between pure science and society.

The iconography of science is rooted in the myth of pure science. The idea that pure science is a purely intellectual or cognitive creation untouched by social facts has been undermined if not yet demolished by sociologists and social theorists from Durkheim and Fleck to contemporary researchers in science studies. How, then, are we to understand what it is that pure science—so often personified in Archimedes, Newton, and Einstein—represents? Let us consider this question in terms of the purest of the pure sciences, pure mathematics.

Pyenson defines pure mathematics as "mathematics pursued for its intrinsic interest, not as a tool in the service of other interests." By introducing the notion of interests, Pyenson shifts our focus from the individual experience of mathematical thought to the politics of pure mathematics.[10] In his *A Mathematician's Apology*, G. H. Hardy wrote:

> I have never done anything "useful." No discovery of mine has made, or is likely to make, directly or indirectly, for good or ill, the least difference to the amenity of the world.[11]

The noted chemist Soddy considered Hardy's views a scandal: "From such cloistered clowning, the world sickens." But J. R. Newman calls Hardy's statement "nonsense"; Hardy's Law is important in the study of Rh-blood groups and the treatment of hemolytic disease in newborns; and his work on Reimann's zeta function has been used in studying furnace temperatures. Hardy's radical defense of purity must be understood as an intellectual strategy. The fact is that Hardy hated war and the application of mathematics to problems in ballistics and aerodynamics. Thus, one aspect of the politics of pure mathematics is that it is an intellectual strategy for responding to social problems, issues, and conflicts.[12]

Within mathematics, the argument that there is a politics of pure mathematics is supported by the perennial rift between pure and applied mathematicians on university faculties. Peano's conflicts with Volterra and other members of the mathematics faculty at the University of Turin are one example of this rift from the early history of professionalized mathematics.[13] The social dynamics of contemporary mathematics are often revealed in these conflicts, which reflect disagreements about how mathematical knowledge should be used and struggles for scarce resources within the university system and in the larger funding arena.

There is no need to deny "the search for knowledge" as an individual or even a collective goal to recognize that the relevance of pure research to societal interests may be something else besides the production of knowledge for its own sake. Pure science may, for example, function as a *demonstration* of the *capacity* for research in a society. Such demonstrations can be the basis for intimidating enemies, projecting status claims, or establishing territorial claims. The centers Germany established for scientific research early in this century in Samoa, Argentina, and China served territorial functions. Today, putting the label "pure science" on the research being carried out in the various national camps and outposts of Antarctica is a way of maintaining informal territorial claims. Because of its generality, pure mathematics plays an important role in establishing the purity of scientific disciplines. One of the few political leaders to acknowledge the political function of pure mathematics was the mathematically inclined Napoleon I, who said that "the advancement and perfection of mathematics are intimately connected with the prosperity of the state."

Purism, then, is an intellectual strategy that has multiple roots and functions. As a *political* strategy, it can demarcate and defend the pursuit of knowledge from military, economic, and political interests one is opposed to; it can be used by ruling elites to establish territorial claims indirectly; and it can help political leaders maintain control over creative and innovative researchers—pure scientists are granted "academic freedom" so long as what they do keeps them from becoming active critics of government or actively interfering with their government's efforts to put their discoveries and inventions to use in the interest of military, economic, or political advances.

Psychologically, purism can be used to satisfy an individual's need for and interest in purity in general as an emotional resource or defense mechanism. This form of purism can become severely pathological if the fear of earthly pleasures and the conflicts of everyday life produce an extreme aversion to anything considered unclean or polluting. An intellectualized purism can develop among "floating intellectuals" who are not committed to or constrained by established social institutions, *but* have failed to develop strong independent ways of establishing for themselves what is true and what is false. This form of purism is associated with weakly formed social, political, philosophical, or religious interests.

Religion and science are often mated in psychological purism and its variations. This is especially the case in mathematics and mathematized sciences. Consistency and completeness, hallmarks of pure mathematics, are central features of the Holy. Pure mathematics and religion were, for example, closely linked in the lives and works of George Boole and W. R. Hamilton. The religious imperative is widely recognized as a feature of early modern science, but its manifestation in contemporary mathematics is not so apparent. Gauss still held to an idea common to his peers and predecessors, that pure science exposes the immortal nucleus of the human soul. Already with Gauss, however, we find a transition from worship of God to worship of Nature as the object of human reason. Gauss still believed in an eternal, just, omniscient, omnipresent God. He was always trying to harmonize mathematical principles with his meditations on the future of the human soul. Cantor believed in the Platonic reality of infinite sets because their reality had, he claimed, been revealed to him by God. And Bourbaki (the pseudonym for an influential group of early twentieth-century mathematicians) claimed that mathematical problems evoke aesthetic and religious emotions.

Ideally, the development of pure mathematics can be portrayed in one of its aspects as a transition from an orientation to and a belief in God, to an orientation to and belief in Nature and then Logic. Indeed, the intuitionist mathematician Brouwer gives a classical Durkheimian analysis of the reification of Logic. He argues that classical logic was abstracted from the mathematics of finite sets and then subsets. This limited origin was then forgotten and logic became viewed as something prior to and above all mathematics. The substitute God, Logic, was then applied without justification to the mathematics of infinite sets. And like God, substitutes such as Logic can serve as *moral* imperatives and constraints that in one way or another—even when we are "left alone" to pursue our own interests guided by our own curiosity to understand the way things work—bind us to established interests at the local professional level and at the level of the state, and reinforce obedience (however irreverent our thoughts and actions are) rather than provoke resistance to and the severing of relationships with established institutions.

The association of modern science with the discipline of the machine (an extension and intensification of the traditional relationship between science and technology) makes any easy assumptions or assertions about modern science as a force for social progress unwarranted. For the very signs we take to mark *material* or *technological* progress, whether in weaponry or medicine, are often indicators of social problems. This helps to explain the highly critical judgments some scientists and science observers have leveled at science. G. H. Hardy, whom I mentioned earlier, defended pure mathematics as strongly as he did as a way of drawing attention to and resisting the fact, as he saw it, that

> a science is said to be useful if its development tends to increase the existing inequalities in the distribution of wealth, or more directly promotes the destruction of human life.[14]

Hardy wrote these words in 1915. About twenty-five years later, J. D. Bernal noted that the image of science was dominated by militarism, economic chaos, and the threat of increasingly terrible wars. Another quarter of a century passed, and Bernal looked out on what he described as "the actuality of a divided world with greater poverty, stupidity, and cruelty than it has ever known."[15] More recently, writers such as Theodore Roszak, Morris Berman, and David Dickson have echoed Hardy and Bernal in their criticisms of science as a world view and a way of life. Eugene Schwartz offers the following representative observation:

> Science and the scientific attitude today are the dominant powers in the world, and they are driving the world toward destruction. . . .
> It is not surprising that among the great driving forces of science are the assault upon nature, the desire for profit, and, above all, the pursuit of war.[16]

Indeed, the association of modern science, violence, and warfare has ancient roots, as David Dickson has noted:

> From its earliest origins in ancient Greece, Western science has enjoyed a close and productive relationship to military power. This relationship has intensified in the forty years since the Second World War, a period in which, building on the experiences of that war, the rapid escalation of military force in both East and West has been grounded increasingly on the applications of advanced scientific knowledge to weapons of mass destruction. Science has done well out of its role, for the rise to positions of influence and favor of the scientific establishment in both hemispheres has been largely due to the contribution science has been able to make to new military technologies.[17]

Feminist theorists such as Carolyn Merchant, Evelyn Fox Keller, Sandra Harding, and Elizabeth Fee have added a new dimension to the critique of modern science by linking it to issues of gender and power.[18]

What the various writers I have cited have helped to establish is that modern science is at the root of our social and environmental ills rather than an unambiguous force for freedom, understanding, and social progress. Their arguments are only convincing, however, if we recognize that the referent for "science" or "modern science" or "Western science" in their writings is *modern science as a social institution.* This is important because "science" can have other referents, including idealistic ones. Because of this multiplicity of referents, many of the critics of modern science experience a dilemma similar to that expressed by Michael Bakunin:

> What I preach then is, to a certain extent, the *revolt of life against science,* or rather against the *government of science,* not to destroy science—that would be high treason to humanity—but to remand it to its place so that it can never leave again.[19]

The dilemma is that on the one hand, it seems intuitively clear or otherwise obvious that "science" in some sense, has contributed to whatever social progress has been achieved on this planet. On the other hand, we entertain no such illusions about *modern science as a social institution*. The dilemma arises in another, related way to the extent that we associate science with *rationality*. Again, objections to rationality, and the claim that rationality has obstructed rather than facilitated social progress, appear to many of us to be unwarranted and perverse, and harbingers of the twin demons of relativism and irrationalism. Indeed, rationality (reason), science, and progress are often viewed as parts of an ideological triad. But just as in the case of science, we find upon closer examination that rationality too is, if not a demon, at least Janus-faced.

THE PROBLEM OF RATIONALITY

In the fourteenth century, the word *rational* meant having the faculty of reasoning, or endowed with reason; and exercising (or being able to exercise) one's reason in a proper manner. In other words, to be rational is to have sound judgment, to be sensible, to be sane. Rational was also opposed to *empirical*. These two words were applied to two classes of ancient physicians who, respectively, deduced their treatment of cases from general principles, and based their methods of practice on the results of observation and experiment rather than "scientific theory." Among the empirical physicians, a practice was adopted if it worked, even if the reasons for its efficacy were unknown. This term was later applied to unscientific physicians or quacks.[20]

By the seventeenth century, the word *rationality* was being used to refer to the quality of possessing reason, and the power of being able to exercise one's reason: the fact of being based on, or agreeable to reason (or, a rational or reasonable view or practice); and the tendency to regard everything from a purely rational point of view.

Today, the lexicographical meaning of rationality is the quality of being rational, which in turn is defined as having reason or understanding. The three basic etymological ingredients of *reason* are: ratio, or proportion; computation (re: to count); and calculate. These terms are associated with and in fact constitute the verb *to think*. *To reason* also has the sense, *to fit*. This meaning is related to the Greek cognate, *arm*, which in turn is associated with *power* and *might*. *To arm* means to furnish with something that strengthens or protects; to fortify morally; to equip or ready for action or operation; and to equip or ready for struggle or resistance. These are the etymological grounds for the standard definitions of *reason:* a statement offered in explanation of justification; a rational ground or motive; a sufficient ground of explanation or logical defense; the thing that makes some fact intelligible (in this sense, reason is a synonym for cause); the power of comprehending, inferring or thinking, especially in orderly, rational ways (here, reason is a synonym for intelligence); the proper exercise of mind (here, reason means sanity). Rationality, finally, means acceptability to reason, reasonableness.

What stands out for the sociologist in the preceding lexicographical, etymological tour is the shadow of community standards that covers the terrain. Words such as *proper, sound, sensible, sane, acceptability,* and *orderly* imply (indeed are synonymous with) prevailing community standards. This is important because one notable aspect of the recurring arguments for and against rationality is that they are implicated in conflicts about community standards, and resistance to the authority of those who set, sustain, and protect community standards—and in the efforts by outsiders to wrestle that authority away from the insiders. There is, naturally, a labeling dimension in these conflicts. The label rational is applied to claims of superiority, the defense of a privileged status or position, and the justification of one's or a group's power to define what is proper, sound, sensible, and sane in a community. The Rational Christians, for example, claimed a superior rationality for their form of Christianity. In this sense, then, rational is not a synonym for science (from the viewpoint of modern scientists and scientific thinkers). From this perspective, every community establishes standards for what is reasonable and labels them rational. This helps account for the adjectification of rationality that gives us scientific rationality, legal rationality, theological rationality, and so on. There may, of course, be overlaps and mutual influences between and among different forms of rationality. But the idea of community-specific rationalities is often used as a way of demarcating science from other modes of thought and knowing. Science is viewed by demarcationists as the only truly rational mode of thought or knowing.

There are two basic forms of opposition to the demarcationist strategy based on the rationality criterion. One is the claim that there are *no* criteria that allow us to distinguish better and worse rationalities across the variety of human cultures, classes, and professions. This claim is often grounded in a misreading of the so-called logic of the Azande made famous by E. Evans-Pritchard's descriptions of their witchcraft and oracular beliefs. The second is the claim that there is a basic set of reasoning principles, or rationalities, that are part of the repertoire of human cultures. Thus, the Azande, plumbers, garage mechanics, the ancient Greek philosophers, paranoids, and physicists all use the same basic principles of reasoning. For example, some anthropologists of science claim that the forms of "laboratory reasoning" are nothing more or less than applications of the unspecific properties of commonsense rationalities identified by Harold Garfinkel; these include:

(1) a concern for making things comparable;
(2) a concern for establishing a "good fit" between observation and interpretation;
(3) a concern for timing, predictability, and correct procedures;
(4) a search for previously successful means;

(5) a conscious analysis of the alternatives and consequences of action;
(6) an interest in the planning of strategies;
(7) an awareness of choices and the grounds upon which these choices can be made.[21]

There are, in other words, no rationalities peculiar to science. The formal features of reasoning show the scientist to be a practical reasoner. Although there is some basis for the second claim, neither claim tells the whole story of rationality as a cultural resource. To begin to fill in the rest of the picture, I turn now to the notion that rationality is a norm of science.

The sociologist of science Bernard Barber identifies rationality as a norm of science: it is "the critical approach to all the phenomena of human existence in the attempt to reduce them to ever more consistent, orderly, and generalized forms of understanding." Barber is one of the few sociologists of science who troubles to articulate and define rationality. Others take it for granted as a norm of science or inherent in science, or ignore it. Robert Merton, for example, says virtually nothing about in his essays.[22]

Philosophers of science have been more concerned with rationality than the sociologists. This is no surprise; the scientific community, like all other complex communities, has a division of labor—the philosophers of science are the moral entrepreneurs of that community. Twentieth-century philosophers have defined rationality in several ways (I follow Laudan here):[23]

(1) acting to maximize one's personal utilities;
(2) believing in, and acting on, only those propositions that we have good grounds for believing to be true (or at least to be more likely than not);
(3) a function of cost-benefit analysis;
(4) putting forward statements that can be refuted.

Now it is easy enough to find cases in the history of science in which most if not all observers would agree "intuitively" that scientists in a given situation were being "rational" even though they failed to follow any standard "model" of rationality. Consider the following ploys of scientists identified by Laudan:[24]

(1) invoke "non-refuting" anomalous problems as major objections to theories
(2) concentrate on clarifying concepts and the reduction of other sorts of conceptual problems;
(3) pursue and investigate "promising" theories, even if those theories are less "adequate" than rival theories;
(4) utilize metaphysical and methodological arguments against and in favor of theories and research traditions;

(5) accept theories even though they are confronted by numerous anomalies;

(6) accept theories which do *not* solve all of the empirical problems of their predecessors;

(7) exhibit wild fluctuations regarding the importance of a problem, and even its status as a problem.

As we will see in a moment, Laudan does not use these examples as a basis for announcing the end of rationality. Let us first consider the arguments of some other critics of rationality in science.

Michael Mahoney has criticized the assumption that scientific *practice* is characterized by rationality. He argues that illogical reasoning is often used in theory evaluation; the only viable means for valid scientific inference available to us appears to be disconfirmation, but disconfirmation is limited in scope and does not actually fit what scientists do in practice. Theories thus seem to be born refuted, because they are virtually always at odds with some data, even when they are first proposed, and (Duhem-Quine) there are no really critical or crucial tests for theories.[25]

In *The Retreat to Commitment,* W. W. Bartley applies the "tu quoque argument"—the "how do you know" argument—to rationalistic (reason-embracing) modes of knowing, and concludes that all such approaches are irrational. They are irrational because they are supported by some sort of ultimate epistemological authority such as logic, revelation, or sense data; and because the authority of the ultimate authority is accepted on faith. Rationality is thus limited by the fact of the infinite regress of standards requiring justification. Because sooner or later we must make a dogmatic commitment, we can in principle make any commitment we want; therefore, Bartley argues, we are not in a position to be criticized for our commitment or to criticize the commitment of others.[26]

Mahoney and Bartley both contend that the effort to justify traditional rationality is at the root of its irrationality. Laudan's critique of traditional rationality, by comparison, is that it involves accepting those statements about the world that we have good reason for believing to be true. Progress, in turn, is usually viewed as a successive attainment of the truth by a process of approximations and self-corrections. He criticizes the argument that actual standards of rational appraisal have remained constant over time. Components of rational appraisal such as ideas about scientific testing, beliefs about inductive inference, and views about what is to count as an explanation have changed enormously. (Note that for Popper and Lakatos, scientific standards of rationality have *evolved;* but the issue is whether a particular theory was well founded by *our* current standards of rationality.) Given the criticisms rationality is subject to, what should we do? Do we jettison or save rationality? Mahoney, Bartley, and Laudan *save* rationality using two different ploys. Mahoney and Bartley argue that rationality should be based on criticism rather than justification. Mahoney's comprehensively critical rationalism (CCR) does not rest on the claim that it is indubitably rational—it thereby avoids the "tu quoque." The CCR strategy is to reduce errors, not guarantee truths.

Laudan's ploy also jettisons justification, but replaces it with a practical approach. He makes rationality parasitic upon progress. Rational choices are progressive choices; they "increase the problem-solving effectiveness of the theories we accept."[27] No assumptions are made about veracity or verisimilitude.

Why should Mahoney, Bartley, Laudan and so many others want to save rationality? And what do their heroics cost them? Mahoney and Bartley end up arguing that logic is one of the basic methods for reducing error via criticism. They claim that logic is not immune to criticism. But certain forms of logic are presupposed in critical arguments. If we reject *modus ponens* and *modus tollens*, for example, we cannot argue in any meaningful sense. This may not be a justificationist strategy; but as we know from such things as the Lewis Principle and DeMorgan's four-valued logic, it is a problematic strategy.[28]

The costs for Laudan are perhaps more subtle. He basically engages in virtuoso philosophical acrobatics in order to save the label "rational" for science. He seems to go out of his way to avoid recognizing that he has a *sociological* problem on his hands, not a *philosophical* one.

What then is all the fuss about? What is at stake in these criticisms and defenses of rationality—and science? What threats are posed by the twin demons, relativism and irrationality? Barnes and Bloor have pointed out that the defense of relativism is in part a reaction to the Cult of Rationalism and the remarkable intensity of that Cult's Faith in Reason (and Science). Earlier, I alluded to the notion that what is at stake here is intellectual turf. Just so, Barnes and Bloor write:

> A plausible hypothesis is that relativism is disliked because so many academics see it as a dampener on their moralizing. A dualist idiom, with its demarcations, contrasts, rankings and evaluations, is easily adapted to the tasks of political propaganda or self-congratulatory polemic. *This* is the enterprise that relativists threaten, not science. . . . If relativism has any appeal at all, it will be to those who wish to engage in that eccentric activity called "disinterested" research.[29]

Rationality is the mantle of those in power, those with authority. When the exercise of power and authority inevitably causes wear and tear on the mantle, the result is the *routinization of rationality,* a process that weakens the adaptive potential of a rationality and makes it susceptible to attack and defeat by a more flexible, unarticulated, unformed, diversified, alchemical system-in-becoming. Such a system in every age in which it emerges from the shadows is labeled relativism and irrationalism, the embryonic rationality of the successful attackers. The recurring conflict between rationality and irrationality—relativism (or science and antiscience) is, from this perspective, a feature of the circulation of authority. Another way to look at this is: rationality formalizes an adaptive strategy; this inevitably means that the adaptive *potential* of the strategy becomes relatively fixed and ultimately declines (again, routinization of rationality). Still another way of looking at this is in Nietzsche's terms: if you are looking for peace and pleasure, then believe; if you are looking for truth, then inquire.[30]

It is interesting to consider whether there has been a change in the way relativism is used within the science studies movement. In the early years of this movement, more than a decade ago now, it seems that relativism was a tool used to criticize the behavioral relevance of the norms of science, including rationality. The relativists were outsiders then, underdogs. As their position has strengthened, as a consequence of the institutionalization of the science studies movement, it seems that relativism has slowly become transformed into a norm of science. Its complete transformation into the new rationality norm of a new generation of "scientists" may only be a matter of time.

Science, rationality, and logic can all be capitalized, routinized, and commodified; adopted as tools and symbols of established (especially state) interests, power, and authority; and ultimately used as barriers to free or open inquiry. During the past fifteen years or so, science studies has helped to revise our conception of science along the lines I have sketched here. But there are limits to what science studies has and can accomplish, and I want to consider these limits along with the positive contributions science studies has made concerning the issues before us.

SCIENCE STUDIES: ITS ROLE IN EXPOSING SCIENCE TO SOCIAL CRITICISM, AND ITS LIMITS

Until the 1970s, the philosophy, sociology, and history of science tended in general to reinforce the idea that modern science is a progressive social enterprise. With the emergence of science studies as a hybrid discipline, modern science came under more critical scrutiny, and scientific knowledge itself became an object of social inquiry. But there is a considerable degree of continuity across the 1970s watershed. One key to this continuity is the myth of the Kuhnian revolution in science studies. The Kuhnians have sustained the traditional uncritical belief in modern science as a well-functioning social system. Thus, in spite of important advances in science studies, it is still widely believed that "normal science" (in Kuhn's sense) is an efficient and productive autonomous research community governed by negotiation; that there are privileged value-free sciences—notably physics and mathematics; and that whatever the social problems of science, they could be solved if only science could be purified or socialized, or if the external contexts within which science operates could be so purified or socialized. But "normal science" is an instance of and a factor in the reproduction of a society burdened by oppressive and alienating work; modern science is intimately coupled to structures of class and power; there are robber barons, entrepreneurs, statespersons, and lobbyists in science who link the interests of science and the state; and conflict is a crucial fuel in the dynamics of science.

It is easy to see some of the basic reasons for the continuity in worldview that links "old" and "new" science studies. The very existence of science studies as a profession is dependent on the goodwill of scientists as respondents and objects of observation and analysis, and the belief among scientists, intellectuals, scholars in general, and the general public that science "works" and produces benefits for society. It is also clear that many of us assume that because "scientific methods" seem to be the only "reasonable" methods to adopt in inquiry, we must also adopt the competitiveness, elitism, alienation, machismo, and other social trappings of modern science (and society). This viewpoint is so deeply imbedded that even when we set out to criticize modern science, we adopt the "scientific approach" with all of its social baggage. And how many of us can afford, psychologically and professionally, to recognize (let alone act on the recognition) that the rationality of modern science is of a piece with the Alice-in-Wonderland rationality of power politics, orthodox economics, and patterns of authority that has infiltrated every sphere of social life.

The new discipline of science studies has helped to uncover important social realities of science based on the seemingly trivial notion that scientists are human and that science is a human activity. We have learned a great deal about such things as the ways in which choosing particular technical assumptions can, to use Brian Martin's phrase, "push an argument." We have deepened our understanding of the nature and significance of selecting, interpreting, and using evidence. We know that in a specialized form of intellectual labor such as science, presuppositions seem to be missing (one reason why science appears to value-free or unquestionably objective) only because they have become built into scientific practice itself. And the more we inquire about why scientific research is carried out, who does it, who can use it, and what it justifies, the more connections to "society" we uncover. And because society in this case is highly stratified, we uncover the social ties between science and the power centers of society. This does not mean that all scientific knowledge is contaminated. As Brian Martin notes,

> Scientific knowledge is not solely the product of the quest for profit or the need to justify war. Rather, scientific knowledge—like the organization of the scientific community and the way scientific research is carried out—is selectively oriented towards these types of ends. In doing research, there are many areas which may be studied. Scientific knowl-

edge is mainly developed in these areas and in those ways which show promise of benefiting powerful groups in society. For example, in electronics, scientific knowledge is organized to help promote communication efficiency (usually one-directional communication) and profit rather than ease of general access and local control.[31]

If we look at discussions of modern science that treat it as a well-functioning and progressive enterprise, we will see that they entail a certain worldview and in particular a theory of social relations. This theory of social relations—which justifies elitism, competition, the alienated activity of "normal science," and the separation of science from ethics and values—is a barrier to social progress. Max Weber describes the rationalization of worldviews as a universal, but above all a European, historical process. Rationalization goes hand in hand with the modernization of the state. In these processes, an other-worldly authority, God, is transformed into a this-worldly authority, Reason as an immanent principle. This carries with it the potential for the separation of rationality and science from ethics.[32]

Researchers in science studies must stop thinking, publicly or privately, of "science" as "physics." The tendency to equate science and physics has obscured the significant discoveries of the social sciences, and made sciences such as ecology, biology, and chemistry second-class modes of inquiry. This equation, along with a psychologistic, ahistorical, and asocial conception of consciousness and behavior, has fueled misguided efforts to link scientific and mystical traditions. This has undermined the potential value of examining alternative cognitive strategies.

We need to pay more attention to the role of ideology in modern science. To say that there is an ideology of modern science means in part that there is a dogmatic support for modern science as a way of life, and a collective cultivation of false consciousness that conceals from scientists the psychological, social, and cultural grounds and consequences of their activities. False consciousness can also manifest itself as a mistaken interpretation of self and social role. The ideology of modern science sustains struggles for power and status, institutional survival, and the use of science (to the extent that it overemphasizes quantification, rigor, control, and prediction) as a resource for reducing personal anxieties and fears.

The pursuit of "science for its own sake" generally requires a commitment to work and profession guided and reinforced by the less enlightened aspects of professionalism (for example, the "publish or perish" imperative, and "grantsmanship"). This makes it difficult to find time for "outside" activities and intensifies the ideological hold of modern (professionalized, bureaucratized) science on scientists and on society. The convergence of the dysfunctions of professionalization and bureaucratization tends to increase specialization and overspecialization in a conflictful division of labor. Occupational and organizational closure (autonomy) increases under these conditions, and creative, critical intelligence, along with the more enlightened motives, is eroded. Ultimately, the ability of people socialized under such conditions to distinguish illusion and reality, hallucinations and material events (or at least to know about these distinctions) is threatened. The final price of runaway professionalization, in conjunction with bureaucratization, and the mechanization of the self, must be first the routinization of rationality, and then the loss of the critical faculties.

A more critical view of science and society in science studies would help bridge the gap between science and values, and help us identify the social conditions that simultaneously facilitate progressive inquiry and social progress.

PROGRESS—AND SOCIAL PROGRESS

The Scientific Revolution made "science," "progress," and "rationality" synonymous. In what is generally recognized as the first modern, secular treatise on the theory of progress, *Digression on the Ancients and the Moderns* (1688), Fontenelle argued that scientific growth represented the clearest, most reliable mark of progress. This relationship between science and progress was expressed in the works of Comte and Spencer. Rousseau, by contrast, argued that "our minds have been corrupted in proportion as the arts and sciences have improved."[33] The twentieth-century version of progress, Schwartz writes,

> turns out to be a blindly hurtling technology that has carried man to the moon, split the atom, created a cornucopia of commodities for a privileged few of the earth, and holds out a promise to carry along with it the remainder of mankind.[34]

It is difficult to sustain the idea of progress in the face of the wide range of problems we are burdened with. The essence of the crisis is that the very forces or production we depend on to mark progress are interlocked with the very problems that make us doubt whether there has been any progress. Camilleri, for example, notes that

> the drug and the mental hospital have become the indispensable lubricating oil and reservicing factory needed to prevent the complete breakdown of the human engine.[35]

It is interesting to view the way in which the optimism rooted in the idea of progress and the idea of science is affected by the unavoidable realities of human experience. For example, about thirty years ago, a panel of distinguished scientists gathered to celebrate—of all things—the centennial of Joseph E. Seagram & Sons, Inc. They were asked to speculate on "The Next

Hundred Years." The idea—or better, the ideology—of science and progress required that the scientists project a positive future. And they did. But what is interesting is the way many of them introduced their speculations. The geneticist and Nobel laureate Herman J. Muller said that the future would be rosy,

> provided that the world does not fall prey to one of the four dangers of our times—war, dictatorship of any kind, overpopulation, or fanaticism.

Harrison Brown prefaced his remarks with the words, "if we survive the next century"; John Weir began, "If man survives." The most bizarre opening sentence was Wernher von Braun's, "I believe the intercontinental ballistic missile is actually merely a humble beginning of much greater things to come."[36]

The idea of *scientific* and *intellectual* progress was fueled by the seventeenth-century advances in science and literature by such cultural giants as Galileo, Newton, Descartes, Moliere, and Racine. The idea of *social* progress was added later. Early in the eighteenth century, the Abbe de Saint Pierre advocated establishing political and eithical academies to promote social progress. Saint Pierre and Turgot influenced the Encyclopedists. It was at this point that social progress became mated to the values of industrialization and incorporated in the ideology of the bourgeoisie.[37] Scientific, intellectual, and social progress were all aspects of the ideology of industrial civilization. But there have been attempts to identify a type of progress that is independent of material or technological progress. Veblen, for example, argued that the various sciences could be distinguished in terms of their proximity to the domain of technology. Thus, the physical sciences were closest to that domain, even integral with it, whereas such areas as political theory and economics were farther afield:

> In the sciences which lie farther afield from the technological domain . . . the effect of the machine discipline may even yet be scarcely appreciable. In such lore as ethics, e.g., or political theory, or even economics, much of the norms of the regime of handicraft still stands over; and very much of the institutional preconceptions of natural rights, associated with the regime of handicrafts in point of genesis, growth and content, is not only still intact in this field of inquiry, but it can scarcely even be claimed that there is ground for serious apprehension of its prospective obsolescence. Indeed, something even more ancient than handicraft and natural rights may be found surviving in good vigor in this "moral" field of inquiry, where tests of authenticity and reality are still sought and found by those who cultivate these lines of inquiry that lie beyond the immediate sweep of the machine's discipline. Even the evolutionary process of cumulative causation as conceived by the adepts of these sciences is infused with a preternatural beneficent trend; so that "evolution" is conceived to mean amelioration or "improvement."[38]

Progress, then, can be viewed in terms of "amelioration" or "improvement" in a social or ethical sense. George Benello argues that

> if we use the term culture in its anthropological sense, there is good basis for saying that primitive South Sea Island cultures are considerably more advanced than our own machine-dominated society. What is implied is that there are certain psychological and ecological universals—laws which define the conditions under which human growth and self-realization can take place, no matter what the level of technology.
>
> Although the material conditions of culture may change and evolve, the basic conditions under which the primacy of the person can be affirmed do not.[39]

I doubt that many anarchists, Marxists, and socialists who have embraced science and technology in their programs for social progress, human emancipation, and individual liberty would agree with the notion that progress based on primacy of the person could be independent of material or technological progress. The very development of the idea of the primacy of the person seems to be dependent on a certain level of social development grounded in scientific and technological advances.

Theodore Roszak, one of the foremost critics of modern science and of the very idea of scientific objectivity, distinguished two types of progress:

> No one who is not lying himself blind to the obvious can help but despair of the well-being that a reductionist science and power-ridden technology can bring. Nothing humanly worthwhile can be achieved within the diminished reality of such a science and technics; nothing whatever. On that level, we "progress" only toward technocratic elitism, affluent alienation, environmental blight, nuclear suicide. Not an iota of the promise of industrialism will then be realized but it will be vastly outweighed by the "necessary evils" attending.
>
> But there is another progress that is not a cheat and a folly; the progress that has always been possible at every moment in time. It goes by many names. St. Bonaventura called it "the journey of the mind to God"; the Buddha called it the eightfold path; Lao tzu call it finding "the Way." The way *back*. To the source from which the adventure of human culture takes its beginning. It is *this* progress which the good society exists to facilitate for all its members.[40]

This is another example, then, of the effort to conceive of progress in terms of ideas about dignity, liberty, integrity, creativity, community-mindedness, and ecological consciousness. My notion of social progress originates in such examples.

Social progress involves an increase in the capacity of human beings individually and collectively to identify, process, store, retrieve, and utilize information and knowledge; it is simultaneously measured by the degree to which a community or society has established the sanctity of human life, the dignity of human beings, and, in Emma Goldman's words, "the right of every human being to liberty and well-being"; and the degree of differentiation in a community or society. Herbert Read writes:

> If the individual is a unit in a corporate mass, his life will be limited, dull and mechanical. If the individual is a unit on his own, with space and potentiality for separate action, then he may be more subject to accident or chance, but at least he can expand and express himself. He can develop in the only real meaning of the word—develop in consciousness of strength, vitality, and joy.[41]

Social progress can move *vertically* to new levels and new ideas, and it can move *horizontally,* to spread new levels and new ideas across more and more of the social landscape. I have been at pains to illustrate why it is difficult to support the idea that "scientific progress" has facilitated or represented "social progress."[42] Whatever positive impact science has had on social progress has been primarily on the horizontal level of ideas.

To the extent that social progress depends on new types and new levels of knowledge about the human condition, it has been facilitated by some aspects of scientific inquiry, even within the institutional boundaries of modern science. But in order to discard the cultural heritage embodied in the term *science,* and to broaden the base of our methodological and theoretical resources, I prefer to work with the more general term *epistemic strategy.* I use the term *human inquiry* for that epistemic strategy or strategies consistent with the idea of social progress. This term may be considered kin to Marx's conception of "human science," his projected mode of inquiry for a communistic society and an alternative to alienated science.[43] In the following section, I adopt a prescriptive perspective on inquiry and social progress.

INQUIRY AND SOCIAL PROGRESS

Perhaps the single most inportant thing we can do to begin to strengthen the relationship between inquiry and social progress is to give the social sciences their due. The sociological dimension in science studies, for example, is not merely a matter of disciplinary politics but part of a Copernican revolution in the social sciences. This revolution, rooted in the works of such social theorists as Godwin, Marx, Weber, Durkheim, and Kropotkin has shifted the individual from the center of the social universe and replaced it with the collectivity. Durkheim, for example, conjectured that social conditions penetrate to the very core of the thinking process itself. Gumplowicz expressed this notion in a particularly strong form:

> What actually thinks within a person is not the individual himself but his social community. The source of his thinking is not within himself but is

to be found in his social environment and in the very social atmosphere he "breathes." His mind is structured, and necessarily so, under the influence of this ever-present social environment, and he cannot think in any other way.[44]

Building on the ideas about collective representations and collective elaborations developed by Durkheim, Gumplowicz, and others, Fleck introduced the concepts "thought collective" and "thought style" in a pioneering study of the genesis and development of the concept of syphilis and the procedure known as the Wasserman reaction:

> If we define "thought collective" as a community of persons mutually exchanging ideas or maintaining intellectual interaction, we will find by implication that it also provides the special "carrier" for the historical development of any field of thought, as well as for any given stock of knowledge and level of culture. This we have designated thought style.[45]

In brief, self and mind are social structures. This discovery is still invisible to most social scientists, let alone to most researchers in science studies. As the nature and implications of this core aspect of the Copernican revolution in the social sciences are realized, we will witness the development of the still very primitive sociological conception of knowledge and cognition that has been emerging in science studies. This will lead to new views of "individual" and "community," and recondition our ideas about the nature of liberty and freedom. This shift is a conceptual, causal, worldview shift. It does not support the subordination of individuals to communities or societies; rather, it provides a sturdier foundation for understanding and realizing the value of the individual (primacy of the person), and the ecological and community bases for individual liberty and freedom. This will follow in part from rescuing the historical origins of sociological perspective from industrial and religious apologists and ideologists such as Saint-Simon and Auguste Comte, and recognizing the contributions of the working classes and of the feminists and anarchists to this revolution in social thought. This will help to ground sociological perspective in a new set of presumptions, especially presumptions about the value of individual liberty, the importance of avoiding investing social agents and agencies with unbridled Authority, and the significance of cooperation and nonviolence as mechanisms for survival and development for individuals and societies. One early result of these developments will be the elimination of the paradoxes and controversies generated by current ideas about the *social construction* of scientific knowledge (reflected, for example, in the relativism-realism controversy).

Modern science is an epistemic strategy as well as a social institution. In general, epistemic strategies develop out of ignorance into configurations of conceptual schemas and theories based on cumulative experience, trial and error, "action in the world" (including various sorts of experimentation strategies). The result is the development of preferred epistemic strategies. Whatever our own preferred strategy, there is a heuristic advantage in critical, comparative studies of epistemic strategies. This provides a solid foundation on which to consider the possibility of a "best possible" epistemic strategy, or mode of inquiry. The "best possible" strategy might, for example, be the one that can incorporate and account for the full range of successes and failures in *all* epistemic strategies past and present. But this suggests a monolithic strategy. It would be better to encourage a range of "best possible" strategies, correlated with a range of dominant utilities such as understanding, technological reliability, intellectual coherence, and truth.[46]

It is conceivable that the range of epistemic strategies could be manifestations of a covering strategy. In any case, best possible strategies characteristically play off tinkering, algorithmic procedures, and more or less formal procedures against one another. Overall, a general tinkering pattern predominates. Of course, experience provides broader and deeper foundations for supporting tinkering. Such foundations, however, should not be considered or treated as monolithic, static, and unchanging, and should not be viewed as logically, ontologically, or epistemologically prior to or independent of the theory and practice of inquiry. Foundations represent "what we know," but should not cast this "what we know" in stone. Foundations should be supportive but not rigid, directive but not dogmatic, well founded but not ultimate. In tinkering, local resources and opportunities are selected and exploited in a "sculpting" of contingencies that lead to different types of social constructions, such as facts, theories, technologies, and myths.[47]

The best possible epistemic strategy can never be a specific strategy such as Western science, Eastern mysticism, quantum physics, or Trobriand Island magic. Demarcationism as practiced by philosophers and others who want to clearly, unequivocally, and once and for all separate "science" from "nonscience," "rationality" from "nonrationality," or "truth" from "falsity" is problematic for reasons rooted in the social nature of inquiry. But this realization does not entail either naive or radical relativism, or the impossibility of some sort of demarcation strategy. It is easy enough to understand why some sociologists, historians, and philosophers of science and knowledge have reached relativistic conclusions by superficially examining the famous Azande "poison oracle" case. Barry Barnes, for example, writes:

We possess no rationality criteria which universally constrain the operation of human reason, and which also discriminate existing belief systems, or their components into rational and irrational groups. Variability in institutional beliefs cannot be explained by a conception of external causes producing deviations from rationality. Likewise, the culture of natural science cannot be distinctive because of its rationality, in a universal rather than a conventional sense.[48]

But this is not a foundation for relativism. We *can* discriminate (demarcate, distinguish) among epistemic strategies based on the scale and scope of past successes, and the probability of future successes (on a larger scale and with increasing scope). There is, in fact, a general strategy found in all "successful" knowledge systems—certainly at the level of societies and cultures. To the degree that an epistemic strategy is in fact successful, to that degree the general strategy I allude to will prevail. This general strategy is distinguished by its *capacity* for criticism, reflexivity, and meta-inquiry. Variations in this capacity distinguish the Azande case from at least some cases of what people have indiscrimately labeled "science." The capacity for criticism, reflexivity, and meta-inquiry is directly proportional to the support for these aspects of inquiry—and the pervasiveness of such inquiry—in the society at large.

More specifically, the best possible epistemic strategy is demarcated from other strategies by the level and degree of development of the *critical* aspects of the various features and contexts of inquiry and everyday life—what Cliff Hooker refers to as the "schema of criticism." Thus, we must be prepared to criticize facts, technological procedures, specific theories, the deepest and the most general levels of our conceptual frameworks, the very manner in which we rank these areas of criticism, and finally the established structures of research and criticism themselves. As we move up through these various levels of criticism, the time intervals for exploring and testing criticisms increases; ultimately, this approach can bring a millennia-long cultural and epistemic tradition into question.[49] This critical enterprise is one that must be built into social structures at all levels—from the social structure of the self and mind to the social structure of a society, culture, or civilization. Democratic and anarchistic social structures generally embody this form of inquiry.

Let us look at this question of inquiry and social structure from another angle. On the one hand, research in the history and sociology of science clearly shows that science is relative to social, political, and economic interests, values, culture, and history. On the other hand, we experience the recalcitrance of the world every day and everywhere in the causal or at least systematic relationships between actions and reactions, acts and consequences. The very fact that I am engaged in this enterprise means that I want to be a realist of some sort, but not the kind of realist who seeks binding necessities and ultimate foundations that can prematurely stifle inquiry or close pathways to information and knowledge. This requires a capacity for not investing conjectures—no matter how self-evident, useful, or timeless they appear—with positive or absolute belief. The trick is to be able to grant the acceptability of *necessary* statements, and the weight of evidence, but to accept all such statements and evidence as no more than "well founded." The anarchist who roots individual liberty in the denial of Authority in all forms is equally unwilling to participate in linguistic institutions such as the institution of Proof.

ANARCHY AND INQUIRY

The social institution of modern science has developed in association with the development of modern industrial, capitalist society. By definition, the institutionalization of modern science involved tying loosely organized pockets of scientific work together, and binding them to state interests. Modern science is a tool of the state; and the state represents and defends with all available means (including the means of violence) the interests of a power elite driven by motives of profit, territoriality, and material gain. This is not the whole story, of course. For even while modern science (along with other modern institutions such as education and religion) serves the state, or more loosely, established interests, it is also—intentionally and unintentionally—a major source of critical and creative energy that is aimed at undermining unbridled state authority and promoting more reasoned ways of solving social, political, and economic problems than now prevail. So, too, there are oases of criticism, resistance, and rebellion within an otherwise conservative educational system designed primarily to feed the industrial and military systems with "machined" humans, and to produce a relatively passive *citizenry* and a relatively active *consumer-ry* willing (unknowingly, for the most part) to trade active, intelligent participation in self-government for the "freedom" to be poor, to go to the shopping mall, to drive the freeways.

Those of us engaged in the dialogue this paper is part of may disagree about the extent to which modern science is a tool of state interests, and the extent to which it is *a,* or *the,* crucial source of criticism and creativity in our society; or about the extent to which modern science is elitist, competitive, and alienative on the one hand, or democratic, cooperative, and non-alienative on the other. But I think we can agree that for modern science to play a facilitative role in social progress, it must be uncoupled from state interests, and from antidemocratic social formations. And the ties between modern science and material or technological progress must be severed, because science in the service of material or technological progress (as we know it from the historical experience of the past three hundred years) cannot serve social progress. I have already noted that these proposals imply a radical transformation of science and society.

The emergence of modern science did not bring "scientific inquiry" into being. However, it did help to focus attention on and articulate ways of inquiry that did not follow party lines, at least in principle. This is reflected in the idea that "disinterestedness" is a norm of science. The problem is that for an epistemic agent to achieve the goal of distinguishing material and non-material events, what is real and what is illusory; to discover how illusions, delusions, and hallucinations are grounded in material events; and to reveal in a constructive way the social foundations of his or her own thoughts, the agent must manage to work "objectively." It turns out that being objective is not a simple matter of *deciding* to be objective, or relying on *intersubjective testing* in any conventional sense. Objectivity is a variable, and a complex social process. Basically, the most objective epistemic agents and communities are those that operate with the most general and diffuse interests. That is, the less one is committed to specific institutional or social interests—nationalism, Catholicism, Buddhism, the scientific bureaucracy, the nuclear power industry—the more objective one's knowledge is. This is the point so often missed when the norm of disinterestedness is discussed; disinterestedness is not rooted in social structures but instead conceived of as something floating about, spiritualized, without its feet on the ground. At the same time, one's distance from conventional and special interests makes one marginal, and creates a communication barrier between interest centers and peripheries. The objectivity generated on the periphery tends to be simultaneously of a higher quality than what is generated at the center, and alien to the needs and perspectives of the center.

What sorts of social formations foster objectivity? The answer, as I have already indicated, is democratic and anarchistic social formations. This is really not so surprising when you consider that the earliest efforts in the sociology of science to answer this sort of question produced the notion that science thrives in democracies, and that communism is a norm of science. What is interesting about the relationship between anarchy (to stay with the term that represents the most advanced stages of democracy and communism) and objectivity is that it clearly and unequivocally makes objective inquiry a consequence of and a condition for social progress. To put it simply, objective inquiry is best pursued in anarchistic social formations, and is itself, at its best, anarchistic. We are more likely to learn things that will promote our survival and enhance the quality of our lives in anarchistic societies and communities than in authoritarian, nonparticipatory social formations. Free Inquiry (to paraphrase a Feyerabend slogan) should always be subordinated to the Free Person in a Free Society, although in practice these form an interrelated mutually reinforcing web of freedoms.

We should be practicing and promoting unfettered inquiry, inquiry unimpeded by Dogma, Authority, and narrowly defined Social Interests, inquiry driven by humane values and a well-developed and pervasive schema of criticism. When we compare competing epistemic strategies, we should ask: What good are they, in what contexts, and for whom; what costs, risks, and benefits do they lead to for individuals, communities, classes, societies, and the ecological foundations of social life?

The anarchist tradition, then, stresses the need to separate inquiry from all forms of unbridled power and authority. Only this separation, and the elimination of the state altogether, can guarantee progressive inquiry *and* social progress. As Godwin argued with regard to the fundamental principle that government is incapable of affording any primary benefits to human beings:

> It is calculated to induce us to lament, not the apathy and indifference, but the inauspicious activity of government. It incites us to look for the moral improvement of the species, not in the multiplying of regulations, but in their repeal. It teaches us, that truth and virtue, like commerce, will then flourish most, when least subjected to the mistaken guardianship of authority and laws. This maxim will rise upon us in its importance, in proportion as we connect it with the numerous departments of political justice to which it will be found to have relation. As fast as it shall be adopted into the practice of mankind, it may be expected to deliver us from a weight, intolerable to mind, and, in the highest degree, hostile to the progress of truth.[50]

My preference for democratic, and ultimately anarchistic, social formations is based on their capacity for decapitalizing "Truth" and giving free rein to skepticism. Nietzsche's remarks on these two aspects of inquiry provide some of the basic ingredients for a philosophy of "nothing matters." The very idea of Truth is, he observes, "conclusive proof that not so much as a start has been made on that disciplining of the intellect and self-overcoming necessary for the discovery of any truth, even the very smallest."[51] Truth in this sense is the province of the "man of conviction":

> *Not* to see many things, not to be impartial in anything, to be party through and through, to view all values from a strict and necessary perspective—this alone is the condition under which such a man exists at all. But he is thereby the antithesis, the *antagonist* of the truthful man—of truth.[52]

"Nothing matters" is, like Feyerabend's "anything goes," a slogan of resistance to established Authority and not an invitation to valueless, undisciplined inquiry. The pervasiveness of skepticism and criticism that in part defines democratic and anarchistic social formations is anathema to Truths and Convictions. The formation of Social Interests is ideally and in principle impossible; for in communities and societies based on cooperation as opposed to those based on more or less deadly competition, social interests—reflections or *antagonistic* interests—fuse with social *goals* and lose their potency as barriers to skeptical, critical inquiry.

In its negative aspect, the anarchist agenda is an offensive against all forms of authoritarianism, mysticism, and supernaturalism. In its positive aspect, it is a defense of and program for freedom and liberty in everyday life and in inquiry. Anarchy always has priority over inquiry—however much these two programs are intertwined. This is a forced choice, brought on by the course of historical and cultural development summarized in my "thug theory of history." Social and cultural change has been driven primarily by greed, profit motives, and the quest for power. As a result, we live in a world dominated by inhuman economics; terrorist, fascist, and authoritarian states; chemical disasters, ecological deterioration, and radiation accidents; the mechanization of selves and commodification of interpersonal relationships; and nuclear winter scenarios and the real possibility of nuclear annihilation. For the most part, this situation is a technologically intensified version of the "normal" human condition—at least in "civilized" societies. A thug theory of history, in terms that are less crude, informs William McNeill's discussion of the emergence of modern European civilization beginning in 1500. His description is not only important as an instance of the thug theory, but also because it focuses on a period pregnant with the Scientific Revolution:

> Europeans of the Atlantic seaboard possessed three talismans of power by 1500 which conferred upon them the command of all the oceans of the world within half a century and permitted the subjugation of the most highly developed regions of the Americas within a single genera-tion. These were: (1) a deep-rooted pugnacity and recklessness operating by means of (2) a complex military technology, most notably in naval matters; and (3) a population inured to a variety of diseases which had long been endemic throughout the Old World ecumene.[53]

McNeill emphasizes the roots of European pugnacity in Bronze Age barbarian societies and the medieval military habits of the merchant classes and "aristocratic and territorial lords of less exalted degree":

> Yet only when one remembers the all but incredible courage, daring, and brutality of Cortez and Pizarro in the Americas, reflects upon the ruthless aggression of Almeida and Albuquerque in the Indian Ocean, and discovers the disdain of even so cultivated a European as Father Matteo Ricci for the civility of the Chinese, does the full force of European warlikeness, when compared with the attitudes and aptitudes of other major civilizations of the earth, become apparent.

McNeill goes on to make a crucial connection, that brings us back to my earliest remarks about the intimate relation between modern science and modern society:

> Supremacy at sea gave a vastly enlarged scope to European warlikeness after 1500. But Europe's maritime superiority was itself the product of a deliberate combination of science and practice, beginning in the com-mercial cities of Italy and coming to fruition in Portugal through the efforts of Prince Henry the Navigator and his successors.

Modern science is a product of these pugnacious, brutal, warlike Europeans. It has served them as a tool, and has consistently been bent to their will. This is one of the grounds for the argument that modern science has a masculine bias.

The anarchists have been in the vanguard of those men, women, and children who have not stood idly by as we have been pushed and shoved by robber barons, pirates, profiteers, bureaucrats, and dictators toward *1984, Animal Farm, Amerika*. The spirit of anarchy and its critique of modern science, technology, and society is present in all who stand with William Morris when he cries:

> What! shall man go on generation after generation gaining fresh command over the powers of nature, gaining more and more luxurious appliances for the comfort of the body, yet generation after generation losing some portion of his natural senses: that is, of his life and soul?[54]

These processes can be reversed—or at least slowed or attenuated—only to the extent that we recognize and act on the affinites that link anarchy and inquiry. Modern science cannot fit into this program—it cannot be a synonym for open inquiry because as a partner in structures of domination and authority over human beings and nature, it has wedded a "tyranny of abstractions" to the tyrannical "rule of men."[55]

CONCLUSION

How we view the problem or question of science and progress depends on whether we are oriented to the interests of science as it is and the prevailing values of our society, or to the prospects for an open (democratic, anarchistic) science in an open society. The critique of modern science and modern society emerges out of the answers to questions rarely posed: what do scientists produce; how do they produce it; what resources do they use and use up; what are the by-products and wastes of their work; what good is what they produce, and for whom; what is the relationship between different types of scientists and different types of publics, clients, and audiences; how do scientists relate to each other, their families and friends, and their colleagues in other professions; what is the relationship between scientists as workers and the owners of the means of scientific production? The hagiography that makes scientists collectively "ingenious," "creative," and "benefactors of humanity" does not tell us what types of human beings scientists are, or what type of world they are helping to build.

In its current institutional incarnation, science is a threat to democracy and an obstacle to anarchy. It is politically and economically aligned with an elite class of military, banking, and corporate leaders who share a narrow view of what constitutes the "national interest" and whose "internationalism" and "global perspective" does not extend beyond considerations of territories and markets. Openness in science and society requires preparing citizens for participatory democracy through education rather than the narrow training that masquerades as education in our schools and universities (amply demonstrated by the standard treatments of religion, politics, economics, sex, and science). Open inquiry means that we can gather information and pursue knowledge without church or state looking over our shoulders. Indeed, the very notion of objectivity depends on inquiry that is guided by broad and diffuse values and interests rather than by the values and interests of specific organizations, institutions, or social classes. The robustness of a mode of inquiry—a science, a rationality, a logic—is measured by the depth and scope of its schema of criticism. This implies that some modes of inquiry are better than others. In other words, we can make a mode of inquiry better than it is and better than its competitors, in part by imbedding in it a failsafe theorem: this mode of inquiry makes no justificatory claims about its own rationality, logic, or scientificity, and nothing in it or its products is to be construed as absolutely true, or absolute or certain in any sense; and in part by adding or enhancing schema of criticism. The structural analogue for these factors in organizations, communities, and societies—and in persons—is necessarily a form of anarchy.

NOTES

1. J. B. S. Haldane, *Daedalus, or Science and the Future* (London: Chatto and Windus, 1923).

2. B. Russell, *Icarus, or the Future of Science* (London: Kegan Paul, Trench, Trübner & Co., 1925), p. 5.

3. H. Karp and S. Restivo, "Ecological Factors in the Emergence of Modern Science", pp. 123–43 in *Comparative Studies in Science and Society,* ed. S. Restivo and C. K. Vanderpool (Columbus, Ohio: C. Merrill, 1974).

4. M. Berman, *The Reenchantment of the World* (New York: Bantam Books, 1984), p. 37; E. Geller, *Thought and Change* (Chicago: University of Chicago Press, 1964), p. 72.

5. T. Veblen, *The Place of Science in Modern Civilization and Other Essays* (New York: The Viking Press, 1919), p. 17.

6. B. Hessen, "The Social and Economic Roots of Newton's 'Principia,'" pp. 151–212 in *Science at the Crossroads,* by N. Bukharin et al. (London: Frank Cass, 1931); G. N. Clark, *Science and Social Welfare in the Age of Newton* (Oxford: Oxford University Press, 1949).

7. R. K. Merton, *Social Theory and Social Structure,* rev. and enlarged (New York: The Free Press, 1968), pp. 661–63.

8. See T. C. Heath, ed., *The Works of Archimedes,* unabridged version of the 1897 edition (New York: Dover, n.d.), pp. xv–xvi. This discussion is based on S. Restivo, *The Social Relations of Physics, Mysticism, and Mathematics* (Dordrecht: D. Reidel, 1983), pp. 251–52.

9. B. Hoffman, *Creator and Rebel: Albert Einstein* (New York: The Viking Press, 1972), p. 221.

10. L. Peyenson, *Neohumanism and the Persistence of Pure Mathematics in Wilhelmian Germany* (Philadelphia: American Philosophical Society, 1983).

11. G. H. Hardy, *A Mathematician's Apology* (Cambridge: Cambridge University Press, 1967), p. 150.

12. J. R. Newman, "Commentary on G. H. Hardy," pp. 2024–26 in *The World of Mathematics,* vol. IV, ed. J. R. Newman (New York: Simon and Schuster, 1956).

13. H. Kennedy, *Peano: Life and Works of Guiseppe Peano* (Dordrecht: D. Reidel, 1980), p. 61.

14. Hardy, *Apology,* p. 120.

15. J. D. Bernal, *The Social Function of Science* (Cambridge, Mass.: MIT Press, 1939); and "After Twenty-five Years", pp. 209–28 in *Society and Science,* ed. M. Goldsmith and A. Mackay (New York: Simon and Schuster, 1964), esp. p. 211.

16. E. Schwartz, *Overskill: The Decline of Technology in Modern Civilization* (New York: Ballantine Books, 1972), p. 31; also see T. Roszak, *Where the Wasteland Ends* (New York: Anchor Books, 1973).

17. D. Dickson, *The New Politics of Science* (New York: Pantheon Books, 1984), p. 107.

18. C. Merchant, *The Death of Nature* (San Francisco: Harper & Row, 1980); S. Harding, *The Science Question in Feminism* (Ithaca: Cornell University Press, 1986): E. Fee, "Whither Feminist Epistemology of Science," paper presented in the conference, "Beyond the Second Sex," University of Pennsylvania, April 1984; E. Fox Keller, *Reflections on Gender and Science* (New Haven: Yale University Press, 1984).

19. M. Bakunin, *God and the State* (1916; reprint, New York: Dover, 1970), p. 59.

20. From the Oxford English Dictionary, and Webster's Collegiate Dictionary.

21. H. Garfinkel, *Studies in Ethnomethodology* (Englewood Cliffs, N.J.: Prentice-Hall, 1967), pp. 272ff. I quote the rendition in K. Knorr-Cetina, *The Manufacture of Knowledge* (New York: Pergamon Press, 1981), pp. 21–22.

22. B. Barber, *Science and the Social Order* (New York: Collier Books, 1952), p. 95; cf. R. K. Merton, *The Sociology of Science* (Chicago: University of Chicago Press, 1973).

23. L. Laudan, *Progress and its Problems* (Berkeley: University of California Press, 1971), pp. 121–22.

24. Ibid., p. 125.

25. M. Mahoney, *Scientist as Subject: The Psychological Imperative* (Cambridge, Mass.: Ballinger Publishing Co., 1976), pp. 140–41.

26. W. W. Bartley, *The Retreat to Commitment* (New York: Alfred A. Knopf, 1962).

27. Laudan, *Progress,* p. 125.

28. The Lewis Principle is that a contradiction entails *any* statement (a theorem in what is sometimes called the axiom system of tautological implication); see B. Barnes and D. Bloor, "Relativism, Rationalism and the Sociology of Knowledge," in *Rationality and Relativism,* ed. M. Hollis and S. Lukes (Oxford: Blackwell, 1982), pp. 44–45, n. 40.

29. Ibid., p. 47, n. 44.

30. F. Nietzsche, *Twilight of the Idols/The Anti-Christ* (New York: Viking Penguin, 1968); *The Gay Science* (New York: Vintage Books, 1974).

31. B. Martin, *The Bias of Science* (O'Connor, Australia: Society for Social Responsibility in Science, 1979), p. 86.

32. M. Winter, "The Explosion of the Circle: Science and Negative Utopia," *Nineteen Eighty-Four: Science Between Utopia and Dystopia,* ed. E. Mendelsohn and H. Nowotny (Dordrecht: D. Reidel, 1984), pp. 80–81.

33. Schwartz, *Overskill*, p. 27.

34. Ibid., p. 28.

35. J. A. Camilleri, *Civilization in Crisis* (Cambridge: Cambridge University Press, 1976), p. 42; and see Berman *Reenchantment*, pp. 7–8.

36. W. E. Moore, ed., *Technology and Social Change* (Chicago: Quadrangle, 1972), pp. 122–30.

37. G. H. Davis, *Technology—Humanism or Nihilism?* (Washington, D.C.: University Press of America, 1981), p. 25.

38. Veblen, *Place of Science*, p. 55.

39. G. Benello, "Wasteland Culture," in *Recent Sociology*, no. 1, ed. H. P. Dreitzel (New York: Macmillan, 1969), pp. 263–64.

40. Roszak, *Wasteland*, p. 426.

41. The quotation on "liberty and well-being" is from Emma Goldman, "My Further Disillusionment With Russia" (1924); and the indented quote is from H. Read, "The Philosophy of Anarchism," cited by M. Bookchin, "Post-Scarcity Anarchism" (1974)—all in G. Woodcock, ed., *The Anarchist Reader* (Glasgow: Fontana, 1977, pp. 161 and 366, respectively.

42. Cf. L. Sklair, *The Sociology of Progress* (London: Routledge and Kegan Paul, 1970), p. xiv, who makes a distinction between *innovational progress* ("progress by means of the production of new things, ideas and process, with maximum *impact* on society") and *noninnovational progress* ("progress by means of the maintenance and diffusion of familiar things, ideas and processes, with minimal *impact* on society"); and see also p. 33. On p. 21, Sklair cites Rousseau, as portrayed in "Dialogue on Progress," by Maurice Cranston, BBC typescript, 1966, pp. 6–7, a dialogue between Diderot and Rousseau: "Science grows up with men's vices. Indeed every science you can name has its roots in some moral defect. Arithmetic springs from avarice, physics from idle curiosity."

43. K. Marx, *Economic and Philosophic Manuscripts of 1844* (Moscow: Foreign Languages Publishing House, 1956), pp. 110–111; *Grundrisse* (New York: Vintage Press, 1973), pp. 699ff.

44. Gumplowicz is quoted in L. Fleck, *Genesis and Development of a Scientific Fact* (Chicago: 1935; reprint, University of Chicago Press, 1979), p. 46.

45. Ibid., p. 39.

46. C. Hooker, "Philosophy and Metaphilosophy of Science: Empiricism, Popperianism and Realism", *Synthese* 32 (1975): 171–231.

47. See the discussion in K. Knorr-Cetina, "The Ethnographic Study of Scientific Work: Towards a Constructivist Interpretation of Science," in *Science Observed*, ed. K. Knorr-Cetina and M. Mulkay (Beverley Hills, Calif. 1983), pp. 115–40.

48. B. Barnes, *Scientific Knowledge and Sociological Theory* (London: Routledge and Kegan Paul, 1974), p. 41.

49. Hooker, "Philosophy."

50. W. Godwin, *Enquiry Concerning Political Justice* (Oxford: Oxford University Press, 1971), pp. 225–226.

51. F. Nietzsche, *Twilight*, pp. 172–73.

52. Ibid., p. 171.

53. W. McNeill, *The Rise of the West* (Chicago: University of Chicago Press, 1963), pp. 569–70 for this and the following quotations.

54. *William Morris, Artist, Writer, Socialist*, 2 vols., ed. by M. Morris (Oxford: Blackwell, 1936), 2: 393; cf. A. Carter, *The Political Theory of Anarchism* (New York: Harper Torchbooks, 1971), pp. 99–100.

55. G. Woodcock, "The Rejection of Politics," in Woodcock, *Anarchist Reader*, p. 136.

Progress, Culture, and the Cleavage of Science from Society

DARYL E. CHUBIN

> . . . scientific progress is not quite what we had taken it to be.
> —Thomas S. Kuhn

> As long as the advancement of science is understood to be
> something other than a sheer drive for power, and something
> more than a mere fueling agent for the engines of military or
> economic nationalism, we will probably achieve a decent bal-
> ance in the ends and the uses of the search for knowledge, or at
> least as far as we can look ahead. It is still worth reminding
> ourselves that in the general scheme of things science pro-
> gresses not by acrobatic leaps, as on its hands and knees.
> —William Carey

INTRODUCTION

It is now axiomatic among social analysts of science that science and
technology create problems for society even as they solve problems on behalf
of society. Yet no concept in the discourse of science is more central than
"progress." Scientists invoke it as an article of faith, their most important
product, and justification for receiving continued support. Although "scientific progress" may be recited like a mantra by researchers, it must have
referents. Measuring the realities that correspond to the rhetoric became the
focus of a project sponsored by the National Science Foundation (NSF) that
a colleague and I at Georgia Tech[1] undertook in 1984. If progress is an
output, then how do we social analysts of science identify it, characterize it,
and use it to understand science?

This paper takes the "scientific progress" project as the springboard for
posing more vexing questions. First, as illuminated by social studies of
science, what is the relationship of scientific progress to social progress?
Second, how do interviews reflect the measurement problems inherent in the
concept of progress? Third, how do the social institutions of science, government, and the media interact to communicate and shape public perceptions

of social progress through technology and medicine? Finally, why is the public's understanding of progress still grounded in stereotypes of science and scientists as marginal to American culture?

The images conjured by science journalists are one step removed from, and therefore dependent on, scientists-as-sources. These sources are self-interested, suspicious, recalcitrant, and egotistical about their work—typical human traits historically omitted from the profile of the scientist. Indeed, this profile has been devoid of humanity and full of caricature; it has changed little in the thirty years since Mead and Metraux[2] first documented the image of the scientist.

Nevertheless, science, media, and the public have grown sophisticated in negotiating meanings of progress. By revisiting scientific progress from the perspective of one now within a federal science policy agency, I advance this proposition: the success of science in waging war and sustaining peace, in promoting health, education, and welfare—all symbols of social progress—*erodes* the autonomy of science, reducing the trust of the public while raising expectations about science. As a corollary, if society demands more *of* science, it also demands greater accountability *from* science. Science becomes a victim of its success. Who is to blame—politicians, journalists, scientists themselves? What can be done to alter public perceptions of science and, for that matter, scientists perceptions of the public?[3]

William Carey[4] has observed that "The nature of scientific progress is no disembodied spirit. It prospers or declines in response to stabilities and discontinuities in the broader environment of which it is an element." In the environment of a democracy, the conduct of science becomes both the problem of and the solution to accountability. Scientific progress and social progress conjoin in the legislative mandate of the National Science Foundation and a body of folk wisdom and practice known as "science policy." In the words of William Blanpied,

> The idea that society actually could and *should* progress seems to have originated sometime between the Renaissance and the Enlightenment at about the same time that modern science emerged. Can one disentangle the concept of social progress from the concept of scientific progress?[5]

Following Shapley and Roy,[6] I contend that such a "disentanglement" was achieved forty years ago when Vannevar Bush offered a vision and rationale for federal support of basic research. With the establishment of NSF, an investment strategy that justified a concentration of human and fiscal resources in the name of scientific progress was born. The technological innovations synonymous with social progress were soon decoupled, relegated to the categories "application," "development," and Weinberg's[7] "external" criteria of scientific choice. The urgency was removed from basic research: it contributes to fundamental knowledge but solves no problem, leads to no immediate invention, and improves the quality of life, if ever, in someone else's lifetime. Today, this wholesale lowering of expectations categorically detaches science of intellectual merit from social relevance. That burden is reserved for research missions such as Apollo, the war on cancer, and *the* Big Science projects of the 1980s (whose initials alone suffice)—SSC and SDI.[8]

Whereas science policy concerns the conversion of knowledge into action (allocations that make a difference), scientists themselves, I will argue, are unconcerned with social progress. Their myopia extends to their personal research program and *its* progress. This cleavage of culture—a federal patronage system that encourages claims of the promise of basic research (shrouded in specialist jargon)—separates science from society and produces the need for accountability. Calls for accountability are society's demand that scientists incorporate a concern for social progress into their research practice, and not just into the rhetoric of that practice. Taxpayers' dollars pay for research; still, few clues to how we measure the returns on this investment[9] can be found on the pages of *Science Indicators* or on the lips of policymakers.

SOCIAL STUDIES OF PROGRESS

What methods do we have to assess what we are getting, and how might we devise new tools or clever methods for doing the job better?[10] The Chubin-Porter project queried scientists "at the bench" and as participants in specialized communities about how far the basic research ideology penetrated. Methodologically, our approach, though it involved straightforward interviewing, by no means assured us that scientists immersed in research would be willing to reflect on the experience. Even if they were willing, would they be able to articulate to nonspecialists—a sociologist and an industrial engineer—what they saw as and meant by "progress"? Terminologically, would their words refer to other concepts that analysts take as synonymous with, or symptomatic of, progress, such as quality, productivity, and discovery? Would their use of "progress" fit with our familiar notions of theory, experiment, or serendipity.

The concept of progress denotes a state or product but it connotes a dynamic process. Social analysts of science thus describe progress as a snapshot of a juncture, episode, or milestone in the social production of knowledge. For example, the act of choosing a problem for investigation, consensus that a replication of an experiment has occurred, the allocation of credit to members of a team, and the transmission (through modeling) of tacit knowledge to neophyte researchers are all signposts on the avenue Scientific Progress. The bibliometrician, ethnographer, and discourse analyst, for example, would each focus on different aspects of the scientist's constructions of, or claims to, progress.[11]

When querying scientists about their perceptions of progress, social analysts lack a technical frame of reference, or what Hagerstrand[12] calls "the 'taken-for-granted' culture of the separate fields of learning . . . elaborated through a mute agreement between those who have been trained in a certain tradition." The social analyst therefore is much like the consumer of popularizations. Cloitre and Shinn[13] find that texts prepared with different audiences in mind (specialist, interspecialist, pedagogical, popular) utilize distinct categories of expository device. Specialist texts are high in referents to experimental protocol, graphic imagery, and quantitative argument. Popularization, in contrast, is high in historical referent, metaphorical imagery, and qualitative argument.

Do scientists, however, talk the way they write? And do they write the way they act? Social analysts have been incredulous about scientists' professed consistency among their talk, texts, and practice. Further, does this lack of consistency among accounts—written, oral, and in-situ observation—reflect differences among scientists, among analysts, or a combination of both? No definitive answers are forthcoming, but sufficient (and often eloquent) doubt has been cast to warrant inquiry into scientists' discourse.[14] And linguists counsel that the study (measurement, understanding) of progress in science is a problem in "functional grammar" that concerns

> how communicative verbal interaction is made possible by the possession of linguistic knowledge, and, conversely, to what extent the organization of natural language is determined by conditions imposed on its communicative use.[15]

Functional grammar brings us full circle to the culture of a science, specifically, the paradigms and tacit knowledge that bind practitioners in an ongoing evaluation of what *they* take to be progressive. In Laudan's[16] terms, progress is embedded in a "research tradition," and, therefore, in scientific practice. Recourse to the scientist's research community overrides categories the social analyst might impose. Instead, we learn to hear the "functional grammar" that the practitioners employ, and to translate their discourse into the grammar of social analysis.[17]

FIRST-PERSON EVALUATION: THE INTERVIEW AS TEXT

In the NSF project, Porter and I examined various types of text: the primary (scholarly) journal literature written for specialists, the pedagogical memoir-honorific essays that reconstruct history, the letter of reference that extols the promise of early research accomplishment, the first-person testimony that places progress into its current research context. In reading these texts, we became sensitive to scientists' use of language in describing research experiences.

We compiled more than sixteen hours of oral testimony from twenty-five scientists and engineers who were sampled purposively across natural science, social science, and engineering disciplines. We considered the interviews exploratory; they were never seen as representative of disciplines, specialties, or sectors of employment. A collective biography of the sample reveals eight chemists/biochemists, and threesomes of neuroscientists, economists, operations researchers/computer scientists, and social scientists. With one exception, all are Ph.D.'s, and most have tenured university appointments.

What did we learn? While interviewing we sensed a contradiction: as social analysts we simultaneously torment the scientists with our ignorance, but flatter them with our interrogation. In all, we identified eight themes that tell different stories. They cluster as: funding; literature, jargon, theory; methodology and instrumentation; commercialization; and ad hominem criticism. Each theme arose in the context of discussing scientific progress and was prominent among the associations researchers made among precursors, conditions, and outcomes.

The expressions of our scientists appeal to incremental advance as the typical mode of progress: an experimental result, accumulation of results, a new synthesis. Yet the interview texts allow for the exciting "eureka" possibility: a breakthrough solution, a reconfigured theory, a brand new line of experimentation. The difference between increment and breakthrough, however, may be one's vantage point: distance (and time) may be needed to see the full implications of change. Looking backward, we may all be Thomas Kuhns; looking ahead, in the laboratories and at the benches, is a more daunting enterprise. The myopia of the here and now produces blurred visions of progress. The glare of urgency and competition of the research front leads to methodological injunction: multiple accounts must be taken, sorted, and compared.

We conclude from these accounts that, first, progress is part of the production cycle of knowledge; this cycle is a process, neither a single product nor a constant state. Thus, central to the culture of a science is an ongoing evaluation of what it takes to be progressive. Second, "technique" is an ascendant characteristic of contemporary science; scientists identify progress with instruments, methodology, and experimental protocol. Third, progress is seen as the sparks that fly *between* researchers; the sparks may explain the growth of interdisciplinary research areas (who, for example, does "surface science" and how does physical chemistry differ from chemical physics?).

COMMUNICATING PROGRESS

The Chubin-Porter interviews illustrate that scientific progress is easy to say but difficult to explain, especially if the interviewee is a scientific researcher and the interviewer is not. Nor is the methodological and terminological dilemma peculiar to social analysts of science. It may be more profoundly felt by science journalists whose job is to "explain" through the translation of functional grammars. The category "media" to which science journalists belong has symbolically assumed another burden as well: the monitoring and reporting that serves as an accountability check on claims that scientists make about progress (or its proxies). As Nelkin writes:

> While scientists and physicians see public communication of medical information as necessary and desirable, they are also aware that it extends accountability beyond the professional community. For once information enters the arena of public discourse it becomes a public resource, opening the way to external control. Thus they seek ways to control the discourse—to influence the images that appear in the press.[18]

The tension between autonomy and accountability is nicely captured by the comments of a science writer: "scientists are to reporters what rats are to scientists. Would scientists allow their subjects to check the interpretation of their behavior?"[19] Simply put, no. Thus, science and the media are two powerful professions each seeking to exert control and extract data from a source.

Elsewhere I have noted Rustum Roy's exhortation that scientists' "tithing" to society begins with explaining their craft to the citizenry (which is precisely what journalists seek). "Such action," I wrote,[20] "builds 'public understanding of science'—a catchphrase often invoked, little understood, and most elusive to measure. Speaking to nonspecialists may be the ultimate maze, which scientists are neither trained to run nor eager to contemplate." The few who are inclined or eager to communicate science to the public, the so-called popularizers, are hardly popular with the scientific community, which sees them as selling out—both literally and figuratively.[21]

That there is a demand for "science news" is reflected in the growth of science sections in major daily newspapers (with a circulation of over 11.5 million).[22] Yet citizens who consume science still represent a minority of the U.S. population, a segment now known as the "attentive public." To educate this public, the popularizer must distill and "dejargonize." Perhaps this is why popularization arouses the jealousy and resentment of most scientists: it trivializes as it simplifies, and demystifies as it reduces the distance between what happens in the laboratory or in the field and its significance for ordinary lives.

Consider this recent exposition on "demarcation" by Carl Sagan in *Parade Magazine:*

> At the borders of science—and sometimes as a carryover from prescientific thinking—are a range of ideas that are appealing, or at least mindboggling: the notion, say, that the Earth's surface is on the inside, not the outside, of a sphere; or the proposition that your soul might return after death as an elephant or a worm; or the conviction that some people have the "psychic" power to bend spoons by looking funny at them. Proponents of these ideas do not much exhibit skeptical habits of thought [which clusters with 'the collective enterprise of creative thinking' in science].[23]

For the public, it would seem, attentiveness spurs science literacy; without it the public cannot be informed. Attentiveness, then, is the *public's* tithing to science. (Parenthetically, all this tithing has not sufficed to generate the advertising revenues to sustain popular science magazines; the best of this genre, *Science 86,* sold out to *Discover* in 1986.)

Communicating progress, as a feature of science, is fraught with professional taboos. For the Carl Sagan–Lewis Thomas–Stephen J. Gould ilk of popularizers, sharing knowledge becomes an obligation in the search for symbolic (and fiscal) support from the culture that pronounced scientific labor virtuous. For most scientists, however, reducing professional distance diminishes expertise and authority, a process comparable to a precious resource being drained. But science is a renewable culture resource; thus, study of the images projected about science and the perceptions that result as a way of knowing assumes heightened importance. A small cadre of researchers trained in science journalism now indeed study "scientists as rats" and the relation of attentiveness, literacy, and both the popular support and cross-examination of science.[24]

Popularizations are quintessentially "trans-scientific" images, to invoke again Alvin Weinberg's map of science, society, and the elastic boundary separating them. What happens within the scientific community (more accurately, within a research community) is regulated by professional norms for making claims; when reporting these claims in the popular media, the autonomy of science becomes subordinated to an exercise of information letting as accountability. Although not all popularizations become the subject of controversy, the closer that claims to scientific progress are identified with advances of near-term societal benefit, the greater the scrutiny (*ex post* and *ex ante*) and ensuing tension that is visible in the public domain.

As revealed in a 1986 survey of Sigma Xi, the scientific research society, a major concern of scientists is the "lack of public understanding of what science is or what scientists do, a condition that leads simultaneously to great expectations and great tredipations—a kind of Dr. Schweitzer versus Dr. Strangelove split.[25] How does journalistic scrutiny level these images?

Too often, I fear, the egotism, defensiveness, and hyperbolic claims of scientists fill the space between expectations and trepidations. This may serve perhaps to clarify public understanding of the scientific issues, but nevertheless reduces trust in the science and technology itself. Nelkin writes:

> Journalists act, in effect, as brokers, framing social reality and shaping public consciousness about biomedical events. . . . Through their style of presentation they lay the foundation for public attitudes and actions. Media coverage of medical events . . . can bring in research funds, and even body parts. . . . Finally, media coverage has implications for political and personal choice.[26]

Technology and Medicine: The Scientific Side of Social Progress

This is admittedly a pessimistic portrayal of the scientist's "tithing process." Good intentions do not necessarily bear desired results: the translation of images into perceptions involves many people, devices, purposes, and a selective cognitive posture to boot. Besides, the public sees almost *no* science per se. It sees new things that work and change the way we think and deal with the world. These belong to the categories "technology" and "medicine"; "science" is just an ingredient—vital, but nebulous and incomprehensible.

Mindful of all this, I offer three brief examples of technology and medicine as the scientific side of social progress—a silver lining that, upon closer examination, contains a cloud. By their coverage alone, these cases have become ongoing controversies. Excerpts of that coverage, and commentary on it, become textual data for the analysis of progress.

CHALLENGER

The explosion of the space shuttle *Challenger* was a kind of culture shock that triggered massive bereavement and then investigation. Nelkin's[27] analysis of media coverage of the disaster underscores the symbolic relationship between nationalism and progress:

> The intensity of journalists' reaction was partly a result of their earlier trust in NASA. The *Challenger* accident brought to light a widespread problem in technical journalism—the tendency to rely on interested sources of information. . . . Reporters for years had accepted what NASA fed them, reproducing the agency's assertions, promoting the prepackaged information they received, and rarely questioned the program's premises or the safety of the operation. . . . For 30 years they had covered the space program as an awesome and pioneering venture, a source of national prestige.

Nelkin concludes:

Years of promotional and uncritical reporting had established the shuttle as a symbol of technological progress and patriotic pride and had obscured the continued risks. . . . The extent of the alliance between these two institutions [NASA and journalism] became obvious when the *Challenger* exploded. And ultimately the promotional publicity backfired, damaging both the credibility of the press and the public's trust in American technological enterprise.[28]

Another view comes from the senior congressional editor of *Aviation Week and Space Technology*. Paul Mann calls it the "sociological story":

Now suppose a reporter had written after the fifth flight . . . that disaster is inevitable (a lot of reporters felt that way) . . . that there is an atmosphere of arrogance and high-handedness; that there's no humility anymore about risk.

If a reporter had said all that before *Challenger* exploded, he would have been denounced, or at best ignored. He would have been scorned by the public as a cynic, an alarmist, a Cassandra, as one more carping journalist tearing things down, especially things that Americans cherish. Faced with 24 successful missions, how could a reporter have been savvy enough to predict *Challenger,* and predict it convincingly?[29]

Mann's point speaks directly to the psychology of progress: as citizens, we want to believe in it and share in its glory, even though its scientific roots are utterly beyond our comprehension. And we do not wish to contemplate the risk attached to the enterprise—the risk of technological failure, loss of life (of pioneers yet), and national grief. We do not perform a psychic impact analysis when it comes to progress. As long as the missions succeed, we are lulled into thinking that "we must be doing it right." Only a setback brings us—excuse the pun—back down to earth. Are the scientists, the popularizers, and the press, then, solely entrusted with the responsibility of naysaying, red-flagging, and restraining the public's enthusiasm for social progress? Our second case supplies a bit more data.

SURROGATE MOTHERHOOD

A "mother" can be thought of as a biological factory, her genes purchased and her womb rented. This is the crass portrait of the surrogate mother—a woman screened and selected to be artificially inseminated with the sperm of an infertile (or diseased) woman's husband. The surrogate is paid for carrying the fetus, then relinquishes the newborn to the biological father. Over six hundred children have been born through such contractual agreements in the past ten years. Clinics broker this third-party gift of parenthood; they advertise for and match surrogates to hopeful couples. When the surrogate refuses to surrender the offspring, however, to whom does the baby belong? Does the surrogate have any legal right? On what grounds does a court decide who shall nurture the baby—her biological or other "mother"?

The Stern-Whitehead case of Baby M in New Jersey broadcast these questions and more as front-page drama, a tug-of-war among psychiatrists and social workers, columnists and lawyers, over the criteria of motherhood and the meaning of informed consent, mental fitness, social class, and life chances. The *New York Times* and the *Washington Post* provided virtually daily coverage. If *Challenger* is a case of culture shock, surrogate motherhood is a case of cultural lag:

> Lacking appropriate laws, the judge in the Baby M case must weigh the sanctity of a contract against the claim of the woman who has given birth. Society has embarked on what has been described as a major experiment without a body of law to serve as a guide.[30]

For now, the biological capacity to procreate through artificial means outpaces our cultural means to assign social roles unambiguously. What is biomedically possible enlarges an ethical and legal grey area. Sorting rights, responsibilities, and interests on a national media canvas assaults our sensibilities: society can award parenthood through decree and change the circumstances of life through custody. We may be in awe of the abstract technology of the feat, but there is anguish in its flesh-and-blood regulation. Is social progress hailed or cursed by such ambivalence? A third case underscores this ambivalence even more starkly.

ORGAN TRANSPLANTATION

The issues and emotions surrounding organ transplants cut right across the U.S. population. Whereas surrogacy is an elective status of very few, organ donation or receipt is a possibility for many.

> A 1985 Gallup poll revealed widespread public support, but considerable ambivalence, about organ donation. Though . . . 75 percent approved of organ donation, only 17 percent said they carried an organ donor card. And while 73 percent said they were "very likely" to give permission to donate a loved one's organs, only 27 percent said the same about their own organs.[31]

The experts—transplant surgeons, organ procurement officers, and bioethicists—debate basic questions: Who lives, who dies, who decides, and who pays? The problem is that there are more patients in need of organs than there are usable "spare parts." The society has not yet devised a system of organ donation that presumes consent and provides a means of transporting harvested organs to the "factory" sites, to use the metaphor again, needing them. Consider these quotes from experts:

> Transplants are a sort of early warning system of the problems involved in allocation of scarce resources.

For the first time in the United States, we have to ask in a big way: How do you ration an expensive medical technology?

And most directly to our current consideration:

> [Organ transplation is] an almost irresistible technology for hospitals because they get such nice publicity from it, and because both doctors and patients feel like second-class citizens if their medical center can't do transplants.[32]

Many insurance companies still refuse to pay for adult liver transplants on the grounds that they are "experimental," even though an NIH Consensus Development Conference in June 1983 declared the procedure a "therapy" and possibly eligible for third-party (Medicare) payment. This has yet to happen, however, so $50,000 to $100,000 in upfront payment grants one eligibility, placing the patient in line for an organ. Even with such affluence, an organ for a terminally ill patient is hard to locate in time. Most die waiting. Public appeals are not uncommon. President Reagan has more than once responded in the media on behalf of the family of a dying child.[33] Society must decide whether the life of a welfare mother, foreign national, drug addict, convicted felon, or alcoholic with a cirrhotic liver is of similar value to that of an infant with congenital liver disease.

Once these issues are faced, then we can confront the morality of moving body parts from a dead person into a living one, or from a living person who can sell an "extra" kidney, for example, to someone lacking a functioning pair. These are political and ethical questions that physicians alone should not decide; such decisions belong to the whole society. But will public understanding increase the donor pool and redistribute the payment burden? And would such reforms and efficiencies constitute a social progress that today masquerades as technological prowess? The medical knowledge and skill that make transplanted organs work does nothing to ensure that the system governing supply and demand is fair. How does the public perceive this dilemma? What can it do to influence policy—a social invention—to prolong or enhance quality of life?

Egotism: The Social Side of Scientific Progress

The three cases hold a mirror up to American society. Do "scientists"—the generic category that includes engineers, medical practitioners, and technicians—object to the reflection they see? Two science writers think so:

> As a reporter, I am neither pro-science nor anti-science. . . . What scientists and doctors do, not what you say about what you do, is ultimately what ought to get reported.[34]

This is more than reminiscent of the methodological injunctions issued by social analysts as a rationale for pursuing scientists' conceptions of their craft and its importance to society. Central to these conceptions is the image not only of modern science, but of the modern scientist. Does the journalist writing in a mass medium like a national newspaper portray the scientist or a caricature? Writing in the *New York Times* about interferon research, Harold Schmeck

> warned his readers not to hope for immediate miracles, [and] a group of scientists complained. They feared that such expressions of doubt in the press would affect their research funding. "It is the public's will and support that make further progress possible."[35]

What is going on here? A decade ago historian George Basalla[36] called it "pop science"—images of the scientist that the media, especially the comics, television, and movies, project. His cogent analysis bridges the two kinds of progress and suggests why, in 1987, the cleavage of science from society persists. Four perceptions contribute to the caricature of scientists.

1. Scientists suffer from an anti-intellectual bias: they are "dangerous because they use their minds to control nature."

2. "Research work done by the scientist is not really 'visible' to the public"; thus the scientist is misunderstood:

> 'They' work hidden from public scrutiny. 'They' speak a language that is incomprehensible to a large segment of the nation, educated or not. . . . Can 'they' be trusted with such knowledge, these men who work in secret and with whom one cannot hold extended discourse?

3. There is "widespread confusion between science and its technological applications":

> Science in popular culture is not presented as an intellectual adventure. . . . Instead, the technological fruits of science are stressed to such a degree that basic scientific research merges with engineering. . . . Science becomes a source of products and power, not a means of helping man better to comprehend himself and his surrooundings.

4. Finally, "by overemphasizing the practical results of his work, especially when seeking public funds," the scientist invites "criticism that might better be directed against engineers, managers, and industrialists."

> By deliberately cultivating the image of the scientist as a remote, superbly dedicated, logical, and humorless individual he creates an easy target for cartoonists and popular satirists. [Consider those] who were scandalized by James Watson's *The Double Helix,* in which scientists

were portrayed with all-too-human foibles and science was depicted as, at times, a haphazard enterprise.[37]

How would we update these perceptions? Surely the exploits of Big Scientists such as the Nobel laureate and high-energy physicist Carlo Rubbia[38] and the neurologist Stanley Prusiner who postulated "prions" as the link between slow viruses and Alzheimer's disease[39] herald a new era of egotism in science. Watson seems tame by comparison!

Likewise, the mystery of laboratory practices has lifted somewhat with the revelations of research misconduct and fraud in science. The role of the media and other documenters of deceit has itself attracted volleys of criticism from those who would prefer to let scientists work as the anonymous "they." Historian Gerald Holton,[40] an extraordinarily sensitive and sensible scholar, delivers an ordinary knee jerk when insisting that most of us may be "pushed into this concern about trimming and forging . . . by excessive and aggressive media exploitation." Broad and Wade's *Betrayers of the Truth*[41] is cited as indicative of the exploitation gripping the scientific profession. Holton seems to scapegoat two science journalists who may not have produced the most compelling book, but who also did not *invent* the research misconduct they report. He then goes on to condemn the aberrant neophytes—Spector, Soman, and Darsee come to mind—as improperly socialized; and he exonerates their mentors—Racker, Felig, and the Braunwald-Hurst pair—as innocent victims of this errant behavior. Is the culture of science—the pressure to win the next grant, publish, and please the paternal lab chief—blameless? Why is there no appeal to the collective ethics of the profession? Why should journalists do the moral accounting for science—and then be lambasted for "sensationalist" reporting?

Lifting the shroud on "normal" and "deviant" scientific practices[42] has brought the fallible scientist to prime-time television, Congress, and the front page. On 11 December 1986, the second-highest-rated series on television, "Family Ties," featured the following theme (I quote from the plot summary in that day's *Washington Post*):

> Alex is crestfallen at the discovery that his favorite economics teacher has falsified information in his struggle to keep up with the university's demands that he publish.

Good triumphed over evil in this half-hour episode; the professor confessed his sin publicly, restoring Alex's faith in the integrity of his mentor. Unresolved, however, was the plot subtheme: the professor's commitment to teaching the next generation, at the expense of a handsome publication record, probably *did* cost him his job. This is more like it. We should applaud as social progress the portrayal of the life and hard times of a working academic scientist.

Eerily, the saga of Walter Stewart and Ned Feder—to publish the results of a content analysis of errors and misrepresentations found in the 109 coauthored papers of the discredited cardiologist John Darsee—was stranger than the fiction of the "Family Ties" episode.[43] For three years Stewart and Feder tried to get their analysis into print while three coauthors of Darsee retained attorneys to threaten legal action if the paper was published. The reputations of their clients would be defamed, so publication was restrained—even though the paper had been accepted for publication in *Nature*.

Frustrated, the authors testified before two Congressional committees, while the *New York Times* and *Science and Government Report* sued the NIH under the Freedom of Information Act to secure some of the lawyers' briefs and get this soap opera of censorship in science before the public.[44] The drama swirling around this case was punctuated on 15 January 1987, when *Nature* published an abridged version of the Stewart and Feder manuscript—a version containing changes not approved by the authors, an editorial decrying the notoriety of the authors' effort, and a reply by Darsee's superior at Harvard Medical School, Eugene Braunwald (whose attorney's briefs dogged the editors of any journal contacted by Stewart and Feder during the three-year ordeal). Braunwald's[45] benign tone belies his vigorous behind-the-scenes lobbying against the "guilt-by-association" manuscript.

This, too, I would call progress: several scientific journals and popular magazines carried news of the ill-fated manuscript. No heroes emerge, but a blow for accountability was felt. The image of science took on a more human face. What plagues the rest of society has now permeated science—malfeasance, lawyers, scandal. This outcome is a somber antidote to what Basalla concluded:

> At the popular level, and even beyond, there lingers remnants of archaic attitudes towards magic and the wizard-alchemist-magician that had been transferred to science and the scientist when they first appeared in history. If this is true, then we must admit the stubborn strength of irrational ideas in a rational society and acknowledge the failure of efforts of science educators and the recognized popularizers of science to act sufficiently effectively at that level of human understanding.[46]

The portrayal alone of science in the media is a measure of social progress. Although the images may be too fixed and blatantly unflattering, they seep into the public consciousness by their sheer repetition. Media coverage that exposes egotism clarifies an important component of progress: those self-assured scientists who will do almost anything to advance their research programs and visions, including becoming 1980s-style pop scientists.

CAN PERSISTENT STEREOTYPES OF SCIENTISTS YIELD SOPHISTICATED PERCEPTIONS OF PROGRESS?

The prevailing images of science—negative, masculine stereotypes—have historically emanated from novelists, satirists, television and movie producers who filled a void in the popular culture. As "marginal men," scientists distanced themselves from the society that gave them sustenance: they were *in*, but not *of*, the culture. They offered few glimpses of their labor and craft and, in the main, fed (by indifference) the caricatures that others projected.

Today the symbols of scientific progress derive more from a science grudgingly responsive to the public. A small cadre of popularizers bear impeccable research, instead of artistic, credentials. But the majority of rank-and-file bench researchers are also attentive to inquiries from the press. They know that public relations is a fine art and that they are presumed to be accessible, if not willing, participants in the storytelling. Why, then, do the negative images of science persist?

Perhaps the sins of autonomy and distance have been replaced by the sins of accountability and self-righteousness. *Challenger,* surrogacy, and transplantation span the full range of cultural expectation: technological tragedy, social ambivalence, and medical wonder. The images are gradually differentiating. As the human emotions of hope and despair are coupled with the technical preoccupations of, say, risk assessment and the prolongation of life, sharper images of science, technology, and medicine penetrate the culture. Though no cause for celebration, this coming out of the laboratory and into the streets[47] signals a changing posture by scientists and the public alike. The scientists admit that progress has its costs, and the public realizes that "magic bullets" and "technological fixes" were always empty promises.

This tempering of images and perceptions stems from organizations such as the Media Resource Center of the Scientists' Institute for Public Information in New York that provides a link between journalists and scientists: the objective is to convey timely and authoritative information in intelligible forms. Scientists also testify before congressional committees, appear on talk shows and as expert witnesses, and lobby like ordinary citizens. Above all, they have become more visible in the news section of the morning paper than in the comics.

This is not to imply that the claims of scientists quoted by the media and in various forums are credible or even accurate.[48] As Robert Cowen, science editor of the *Christian Science Monitor,* asks:

> What is truth, what is objective, and what is professional? Certainly, it's not professional for us simply to accept what we're told. . . . Certainly, it's not untruthful to report what we are told at scientific symposia and at press conferences.[49]

Scientists can be as self-serving as any information source. But it is refreshing to witness the same scrutiny and skepticism routinely directed to spokespersons for *other* social institutions applied to science and technology. But whereas I am more sanguine about the *reporting* of science (in amount and diversity), Nelkin is less sanguine about the *kind* of reporting (in tone and intent):

> Many journalists are, in effect, retailing science and medicine more than investigating them, identifying with their sources more than challenging them.
> If the popular press is to mediate between medicine and the public, to help the public to make complex judgments about crucial policy issues, and enhance the credibility of medical and technological choices, both journalists and the medical community must come to terms with the different mode of communication. . . . To understand modern scientific medicine, readers need to know its context: the political and economic bases of decisions, the social and ethical implications of research, and the limits as well as the power of science and technology as applied to problems of health.[50]

The only science journalist working today who consistently displays what Nelkin prescribes is publisher and editor Dan Greenberg. His brand of vigilance, however, is apparently in short supply. Consider this sample Greenberg quip:

> I'm sure the chemists have many grievances regarding the coverage they've been given in the press. My perception and my feeling toward this is that the press, if anything, is too light, too easy on industry as well as on any other institutions that it covers. My experience in covering stories involving industry and government agencies is that anything short of adulation is regarded as hostility.[51]

The Nelkin-Greenberg prescription (with roots traceable to Goodfield and, for that matter, Bernal) is for a profession of science critics, or what I would call "trans-scientists," to evaluate expert claims for public consumption. (Such a profession would require an amalgam of talents: reflexive social science, muckraking journalism, and unusual public policy.) Those claims are no longer couched in the simple rhetoric of basic research; that will not do. With public patronage at stake, the claims are bolder; they demand more sophisticated interpretations that constitute a political economy of the science in question.

Yet by speaking to bench researchers about scientific progress, I learned that the images of their own work and its place in the culture are at best diffuse and ill-formed. It has many referents and, for the nonspecialist, an impenetrable functional grammar. Today's wizards are rather inarticulate about "scientific progress"; they want to be known by their deeds. Those who translate deeds into the discourse of the science literate embrace products and applications as symbols of "social progress." But they, as any reporter or analyst who surveys the intersected region between science and society, must beware of outdated symbols. The discourse must change—and we can help.

CONCLUSION

Science is an ongoing process, an extension of the ego, not just a collection of artifacts that can be bundled, labeled, and delivered periodically to a waiting public as "progress." That is why progress is such an elusive concept—rhetorical at its core, thing-oriented in its measure. Because scientists and journalists are symbolic partners, their mutual need to negotiate the concept and refine its significance will not diminish the tension that exists between them; it will only intensify. But out of this tension will come new definitions of progress that blur the distinction between the scientific and the social, and erode the stereotypes and parochialisms sustained both by journalist and scientist.

Will bottom-line allusions to missions, cures, and panaceas ever recede from our texts and talks? Will the conceptualization of progress routinely encompass the public interest? Only if the autonomy of science continues to erode along with the stereotypes that envelope it. For this to occur, science policy must gain a new bearing, acting on the refined image wrought by the tug-of-war between journalist and scientist, while narrowing the cleavage between science and society.

NOTES

This essay was prepared for the Regional Colloquium in Technology Studies on Science and Social Progress, Lehigh University, Bethlehem, Pennsylvania, 27 February 1987. The views expressed are the author's alone and do not reflect those of any organization—past or present—that has supported his work. Conversations with Lisa Heinz and commentary by Steven Goldman contributed mightily to this revision; alas, we may all yet regret that I ignored some of their sage advice.

1. Daryl E. Chubin and Alan L. Porter, *Measuring Scientific Output: A Collective Biography Approach,* Interim Report to the Division of Policy Research and Analysis (Atlanta, Ga.: National Science Foundation, February 1985); Daryl E. Chubin and Alan L. Porter, *Measuring Scientific Output: A Collective Biography Approach,* Final Report to the National Science Foundation, Grant PRA84-13060 (Atlanta, Ga.: Georgia Institute of Technology, August 1986).

2. Margaret Mead and Rhoda Metraux, "Image of the Scientist among High-School Students," *Science* 126 (30 August 1957): 384–90.

3. Kenneth Prewitt, "The Public and Science Policy," *Science, Technology, and Human Values* 7 (Spring 1982): 5–14.

4. William D. Carey, "Science and Public Policy," *Science, Technology, and Human Values* 10 (Winter 1985): 7–16, quotation from 12.

5. In Arnold Thackray, "The Historian and the Progress of Science," *Science, Technology, and Human Values* 10 (Winter 1985): 17–27, quotation from 23.

6. Deborah Shapley and Rustum Roy, *Lost at the Frontier* (Philadelphia: ISI Press, 1984).

7. Alvin Weinberg, "Impact of Large-scale Science on the United States," *Minerva* 134 (1961): 161–64.

8. Daryl E. Chubin, "Research Missions and the Public: The Over-Selling and Buying of the U.S. War on Cancer," in *Citizen Participation in Science Policy*, ed. J. Petersen (Amherst: University of Massachusetts Press, 1984), pp. 109–29.

9. Office of Technology Assessment, *Research Funding as an Investment: Can We Measure the Returns?* A Technical Memorandum (Washington, D.C.: Government Printing Office, April 1986).

10. Daryl E. Chubin, "Scientific Progress: An Interim Report." *BioScience* 36 (April 1986): 234–35.

11. Daryl E. Chubin and Sal Restivo, "The 'Mooting' of Science Studies: Research Programs and Science Policy," in *Science Observed*, ed. K. D. Knorr-Cetina and M. Mulkay (London: Sage, 1983), pp. 53–83.

12. T. Hagerstrand, ed., *The Identification of Progress in Learning* (Cambridge: Cambridge University Press, 1985), quotation from ix.

13. M. Cloitre and Terry Shinn, "Expository Practice: Social, Cognitive, and Epistemological Linkage," in *Expository Science: Forms and Functions of Popularization,* Sociology of the Sciences, vol. 9, ed. T. Shinn and R. Whitley (Dordrecht: D. Reidel, 1985), pp. 31–60, quotation from 34.

14. For example, Bruno Latour, "Give Me a Laboratory and I Will Raise the World," in Knorr-Cetina and Mulkay, *Science Observed*, pp. 141–70; M. Mulkay, J. Potter, and S. Yearly, "Why an Analysis of Scientific Discourse is Needed," in Knorr-Cetina and Mulkay, *Science Observed*, pp. 171–204; Michael Mulkay, "The Scientist Talks Back: A One-act Play, with a Moral, about Replication in Science and Reflexivity in Sociology," *Social Studies of Science* 14 (1984): 265–84.

15. S. C. Dik, "Progress in Linguistics," in *The Identification of Progress in Learning,* ed. Hagerstrand (Cambridge: Cambridge University Press, 1985), pp. 115–42, quotation from 123.

16. Larry Laudan, *Progress and Its Problems: Towards a Theory of Scientific Growth* (Berkeley: University of California Press, 1977).

17. Michael Mulkay, "Norms and Ideology in Science," *Social Science Information* 15 (1976): 637–56; Trevor Pinch, "Theory Testing in Science," *Philosophy of the Social Sciences* 15 (1985): 167–87.

18. Dorothy Nelkin, "How to Doctor the Media," *New Scientist,* 20 November 1986, pp. 51–56, quotation from 54.

19. In June Goodfield, *Reflections on Science and the Media* (Washington, D.C.: AAAS, 1981), quotation from 94.

20. Daryl E. Chubin, "Scientists as Rats," *BioScience* 36 (June 1986): 357.

21. Rae Goodell, "Problems with the Press: Who's Responsible?" *BioScience* 35 (1985): 151–57.

22. Barbara J. Culliton, "Science Sections in U.S. Newspapers Increase Dramatically in Past 2 Years," *Science* 235 (23 January 1987): 429; "Science gains in newspapers," *Science and Government Report,* 15 January 1987, p. 7.

23. Carl Sagan, "The Fine Art of Baloney Detection: How Not to be Fooled." *Parade Magazine,* 1 February 1987, pp. 10–11.

24. Sharon M. Friedman, Sharon Dunwoody, and Carol L. Rogers, *Scientists and Journalists: Reporting Science as News* (New York: Free Press, 1986).

25. Jack Sommer, "Big Three, Little Four," *American Scientist* 75 (January–February 1987): 112, 103, quotation from 103.

26. Nelkin, "Media," p. 56.

27. Dorothy Nelkin, *"Challenger"* The High Cost of Hype," *Bulletin of the Atomic Scientists* (November 1986): 16–18, quotation from 16.

28. Ibid., pp. 17–18.

29. Paul Mann, "The NASA Story We Missed," *Bulletin of the Atomic Scientists* (November 1986): 18.

30. E. Kolbert, "Baby M Adds Urgency to Search for Equitable Laws," *New York Times,* 15 February 1987, p. E22.

31. Don Colburn, "Transplants: Who Lives? Who Decides?" *Washington Post Health,* 20 January 1987, pp. 14–16, 18, quotation from 16.

32. Ibid., pp. 14–15.

33. Gerald E. Markle, and Daryl E. Chubin, "Consensus Development in Biomedicine: The Liver Transplant Controversy," *The Milbank Quarterly* 65 (Winter 1987): 1–24.

34. Earl Ubell in Nelkin, "Media," pp. 54–55.

35. Ibid., p. 54.

36. George Basalla, "Pop Science: The Depiction of Science in Popular Culture," in *Science and Its Public: The Changing Relationship,* ed. G. Holton and W. A. Blanpied (Dordrecht, Holland: D. Reidel, 1976), pp. 261–78.

37. Ibid., pp. 270–73.

38. M. D. Lemonick, "How to Win a Nobel Prize," *Time,* 9 February 1987, p. 55.

39. Gary Taubes, "The Game of the Name is Fame. But Is It science?" *Discover* (December 1986): 28–31.

40. Gerald Holton, "Niels Bohr and the Integrity of Science," *American Scientist* 74 (May–June 1986): 237–43.

41. William Broad and Nicholas Wade, *Betrayers of the Truth: Fraud and Deceit in the Hall of Science* (New York: Simon & Schuster, 1982).

42. Daryl E. Chubin, "Research Malpractice," *BioScience* 35 (February 1985): 80–89.

43. Daryl E. Chubin, "A Soap Opera for Science," *BioScience* 36 (April 1987): 259–61.

44. Tabitha Powledge, "Stewart-Feder (Finally) in Print," *The Scientist,* 9 February 1987, p. 12.

45. Eugene Braunwald, "On Analyzing Scientific Fraud," *Nature* 325 (15 January 1987): 215–16.

46. Basalla, "Pop Science," p. 275.

47. Twentieth Century Fund, *Science in the Streets* (New York: Priority Press, 1984).

48. E. J. Imwinkelried, "Science Takes the Stand: The Growing Misuse of Expert Testimony," *The Sciences* (November–December 1986): 20–25.

49. In J. Krieger, "Panel Addresses Chemistry's Public Image," *Chemical & Engineering News* 64 (29 September 1986): 28–30, quotation from 29.

50. Nelkin, "Media," p. 56.

51. In Krieger, "Panel," p. 30.

Scientific Progress and Models of Justification: A Case in Hydrogeology

KRISTIN SHRADER-FRECHETTE

INTRODUCTION

No one, as yet, has been able to give a wholly satisfactory account of scientific progress.[1] From the 1930s through the 1950s, most philosophers of science believed that scientific progress and scientific consensus were based on rules of evidence and theory choice; as Laudan puts it, they believed that there was a set of algorithms that would permit any impartial observer to judge the degree to which a certain body of data rendered different explanations of those data true or false, probable or improbable.[2]

Despite the attempts of the logical empiricists to protect the search for scientific truth, we have been unable to arrive at the algorithms believed necessary to safeguard scientific objectivity. Hence, at least by the 1960s, most philosophers of science concluded that the logical empiricists' hopes for scientific progress and consensus were probably not capable of realization. Writers such as Kuhn,[3] Feyerabend,[4] and Mitroff[5] emphasized scientific controversy, the alleged incommensurability among theories, and the problems that methodological rules typically underdetermine a choice among theoretical judgments, and that factual claims often underdetermine a choice among methodological rules.

As Laudan points out,[6] authors such as Popper and Reichenbach believed that cognitive goals and values (e.g., simplicity, heuristic power) could enable one to choose among alternative methodological rules.[7] Hence, even though methodological rules might underdetermine a choice about factual claims, one could appeal to cognitive goals and values to settle disputes about methodological claims. In this way, one could follow a hierarchical model of justification, from (1) factual and theoretical claims, up to (2) methodological rules, up to (3) judgments about cognitive aims or goals; according to this view, one could resolve disagreements at one level of the hierarchy simply by appealing to the next higher level. Hence, methodological rules might help adjudicate conflicts over the facts, and cognitive goals might help resolve methodological disagreements.

Appeal to cognitive goals and values, however, did not enable the logical empiricists' hierarchical model to rescue their notion of scientific progress. For, in their important attempts to guarantee the value-neutrality of science, they denied that judgments about cognitive values and goals had any place in science. They barred such judgments from science because they said that adoption or amendment of a cognitive goal or value is an emotive or subjective matter, not one for rational resolution.[8]

Those who attacked the hierarchical model of the logical empiricists likewise offered no hope of scientific progress. Although Kuhn drew our attention to the role of cognitive standards and values in science, members of both the Kuhn-Feyerabend camp and the positivist camp claimed that cognitive values are immune to rational debate.[9] But if they are, and if cognitive values are central to science, then any hope of a rational account of scientific progress is impossible. And if it is impossible, then both science and society are in trouble. As Scheffler noted, without an ideal of scientific progress, dependent upon a notion of scientific objectivity, then our daily decisions, in both private and public-policy arenas, cannot be based on reason but only on personal authority, partisanship, and coercion.[10]

In this essay I argue that, at least in some cases, a rational account of scientific progress is possible. I claim that often we can rationally assess cognitive values. To establish this point, I shall use a case study in hydrogeology to show that faulty inferences about methodological rules and cognitive values were responsible for erroneous scientific conclusions; that rational and impartial judges could probably agree on how to adjudicate these hydrogeological disputes over methods and goals; and that, if a rational justification procedure were used, in this case, then the faulty hydrogeological conclusions could have been avoided. Next I shall examine this hydrogeological case in order to determine what it tells us about instances in which rational negotiation about cognitive values is possible. I argue that the case suggests that, in virtually all problems of *applied science,* it is possible to adjudicate conflicts over cognitive values, in large part because we are able to employ a fourth type of justification, appealing to ethical or utilitarian goals.

With Laudan, I believe that it is best not to conceive of scientific justification in terms of the hierarchical model of the logical empiricists. This is because in science, we do not merely use cognitive goals to choose among methodological rules and methodological rules to choose among factual claims. Rather, our factual claims influence the sorts of methodological rules that we use, and our methodological rules help determine the kinds of cognitive goals that we choose. In other words, one of the key insights of Laudan's triadic model is correct, viz., that it is reticulated, or that cognitive goals, methodological rules, and factual claims all influence each other; the process is not primarily unidirectional and hierarchical, moving from factual claims up to cognitive goals. Although I agree with the basic correctness of Laudan's model, I wish to add to it another parameter, ethical goals. On this new or quadrilateral model, I shall show how we can adjudicate factual claims by appeal to methodological rules, we can evaluate disputes over methodological rules by appeal to cognitive goals, and we can assess controversies over cognitive goals by appeal to ethical values. Nevertheless, factual claims, methodological rules, cognitive goals, and ethical values all affect each other and are evaluated in terms of each other, especially when one is adjudicating conflicts in applied science. Even in some instances of nonapplied or *pure science,* I shall argue that the appeal to a fourth type of justification, in terms of ethical or utilitarian goals, can be used to adjudicate disputes over cognitive goals or values.

If my analysis is correct, then the logical empiricists' hierarchical model errs in appealing to *levels* of justification. Laudan's triadic model, while correctly avoiding an analysis in terms of levels, errs through incompleteness, because it recognizes only three classes of judgments (those about facts, about methods, and about cognitive goals) for resolving disputed scientific claims. If my analysis is correct then, at least for some clearly defined cases, we need a four-parameter, nonhierarchical model, one including ethical values, for resolving scientific controversy over cognitive goals. To see why this conclusion follows, let us begin by examining the role of value judgments in science.

THE ROLE OF VALUES IN SCIENCE

If science is unalterably value-laden, and values are at least in part subjective, then saving science from subjectivity or irrationalism, and giving an account of scientific progress, requires that the philosopher must accomplish at least two tasks, one theoretical and the other practical. The *theoretical task* is to provide a philosophically satisfactory account of the relationship between facts and values, so that we can establish the precise ways in which values, especially as used in science, are in part objective or factual. It will not do to content ourselves with the vague dictum that all facts are value-laden; if we want a rational account of science as a value-laden enterprise, then we must provide a foundation for the rational evaluation of values. The *practical* task is to arrive at some explicit criteria for making value judgments in science. Let's attempt the theoretical analysis first, in hopes that it will inform the later practical task.

SOLVING THE THEORETICAL PUZZLE ABOUT FACTS AND VALUES

Several philosophers have attempted to sort out the difficult theoretical problems surrounding the fact-value relationship, in order to shed some light on the role of values in science. In his *Science and the Theory of Value,* Peter Caws provides one strategy for rescuing values from the realm of subjectivity. In effect, he makes facts and values species of the same genus, facts, and then explains that what we call "facts" are "retrospective facts," and what we call "values" are "prospective facts."[11] In thus making values "prospective facts," Caws moves values from the realm of mere subjective attitudes and states of mind to the realm of future states of the world, and hence to the set of things about which, in principle, we might have evidence or seek confirmation. Caws's exact words are: "A value is a future fact, selected from among a set of alternative, and mutually exclusive, future facts, and marked with an imperative."[12]

Now what makes Caws's account of value so compelling is not only that, if he is correct, he has rescued values from the realm of subjectivity, but also that his account coheres, in certain respects, with the way in which we seem to speak of values. For one thing, Caws's account focuses on concrete embodiments of values, future facts, and not merely on values as abstract qualities—for example, pleasure. He speaks of "pleasureable experiences" as values because we hope certain pleasureable experiences will occur in the future. If they do not, then they are unrealized values. Likewise, on Caws's view, one might speak not of the abstract value, consistency, but of "consistent experimental results" as values because we hope to obtain future experimental results that are consistent with the test implications of our current hypotheses.

There are a number of problems with Caws's account of values as future facts. These problems seem so severe that they appear to undercut any hope that his scheme is capable of delivering values from subjectivity. *First,* in calling values "future facts," Caws appears to confuse future facts with characteristics of future facts. If future experimental results are consistent with one's hypothesis, and one's value of consistency is realized, for example, then consistency is a *characteristic* of the future fact, a characteristic of the experimental results, not the future fact itself. Now Caws claims that the values are the future facts, and not characteristics of them, because we would not know that the abstract qualities, like consistency, were values if we had not first attached value to their concrete embodiments. Caws's claim here errs, however, because it confuses how we know values (in their concrete embodiments) with what values are (as characteristics of concrete things). Caws confuses epistemological justification with ontological meaning.

To say facts are future values not only confuses how we know them with what they are, but also appears to involve a category mistake. It seems incorrect to claim that what we know as facts and values are really only species of the same genus, facts. Indeed, the differences between the two alleged species are so great that it is difficult to see how they could both be members of the same category or genus. Perhaps the chief of these differences is that one confirms facts by observation, and one substantiates values by argumentation. Therefore, just as Ryle rightly criticized the person who believed that a university could be identified with buildings, so also one might criticize Caws, who seems to believe that values can be identified with facts. It seems just as peculiar to say that "this value is a future fact" as it does to say that "this university is a future building."

Admittedly, it may be true that values are facts, either in the Platonic sense that the highest values reside in real objects, forms, or in the social-scientific sense that values, in the sense of preferences, are facts.[13] Even if values are facts in either of these two senses, however, it must also be the case that values are characteristics of facts, not merely facts. Moreover, even if values are both facts and characteristics of facts, values are also characteristics of our accounts of facts, characteristics of our theories. Caws admits that we speak of values such as pleasurable experiences and honorable actions. But it appears also that we speak of values such as consistent theories or simple explanations. And if so, then values are characteristics of our theories, as well as characteristics of our facts, as well (perhaps) as facts themselves.

Second, if Caws were correct in alleging that values were merely future facts, then values and facts would be confirmed in the same way, by evidence. Yet Caws himself admits that facts and values are not confirmed in the same way. He says that facts are confirmed by asking "whether such and such phenomena are in fact observed,"[14] whereas values are validated by asking "not only whether such a world really would result, but also whether this is the kind of world we want."[15] But if facts are confirmed by asking whether phenomena exist, and if values are confirmed by asking whether we want them to exist, then it does not seem consistent to claim that values are merely one type of fact, prospective or future facts.

Third, Caws also seems to err in calling values "future" facts when, at best, he appears to mean they are "possible" facts. He admits that some future facts (values) "may not come about" and hence are unrealized values.[16] But it seems that facts that do not come about are not future facts, but possible facts. Caws defends his calling values "future" facts by claiming that the label, "possible" facts, will not do, because "people may very well attach value to impossible facts."[17] His defense seems questionable, however, both because he gives no examples of "impossible facts," and because all facts appear to be possible.

TYPES OF VALUE JUDGMENTS IN SCIENCE

If Caws's account of values is not successful, then it is at least in part because value, goodness, or worth is close to a primitive concept, and therefore hard to specify, and because what we call "values" are so multi-faceted. They appear, in quite different senses, to be both facts and possible facts, and to be both characteristics of our facts and characteristics of our theories. Moreover, facts and values are not mutually exclusive, so any account of them and their relationship is unlikely to be based on clean, and therefore philosophically satisfying, distinctions. Because of these difficulties with analyzing the concept of "value," in itself, it might be more instructive simply to attempt to specify the different types of values, and then to determine possible criteria for judgments about each of the types, depending on their relevance to science.

Longino, for example, distinguishes three different types of values, bias values, constitutive values, and contextual values, all relevant to science.[18] *Bias values,* on her account, are deliberate misinterpretations and omissions to serve one's own purposes; these are the sort of values from which science ought to be free. *Constitutive values* are those underlying particular rules of scientific method; presumably these include factors such as simplicity and heuristic power. *Contextual values* are personal, social, cultural, or philo-sophical emphases and goals; one example of a contextual value would be commitment to an atomistic, rather than field-theoretic, framework for high-energy physics research.

The Longino values-scheme is useful because it distinguishes different types of values that arise in science and clarifies which type (bias values) ought and can be avoided, as the logical empiricists emphasized, and which types (contextual and constitutive values) are impossible to avoid, as Kuhn and others emphasized. Her account explains how both the Carnapian pro-ponents of value-free science[19] and the proponents of postpositivist, value-laden science[20] are both correct, but in different respects. The deficiencies of the scheme, however, are that the three categories of value are not mutually exclusive and hence do not suggest clear criteria for making judgments in science. Bias values are often inextricably connected with constitutive values, for example, especially in cases of scientific judgment that involve the absence of (what Feynman calls) a "smoking gun." Moreover, a great variety of quite different values appears to be included under the category of con-stitutive values; to arrive at criteria for judgments about constitutive values, one would probably need to distinguish the various types.

McMullin does a more precise job of distinguishing different types of constitutive values. He begins by noting that values can be either emotive (cf. Longino's "bias values") or pragmatic (a result of finite time or resources) or cognitive, and that only the former have no clear place in science. Within cognitive, epistemic value judgments (cf. Longino's "constitutive values"), he distinguishes *evaluating* from *valuing.*

On McMullin's analysis, the key to understanding evaluating and valuing is what he calls *characteristic* (epistemic) values, objective properties of things. For example, speed is a characteristic value of sports cars. Judgments involv-ing characteristic, epistemic values arise in two ways, says McMullin. When we evaluate the extent to which a particular thing possesses a characteristic value, for example, when we evaluate how fast a sports car will go, we may be said to be *evaluating,* making a largely *factual* judgment regarding characteristic value. But when we assess the extent to which a characteristic, such as speed, is really a value for the thing (the car) in question, then we may be said to be *valuing,* making a largely *subjective* judgment regarding the characteristic value of speed.[21]

Hempel made a distinction quite similar to McMullin's when he spoke about instrumental, as opposed to categorical, judgments of value. For Hempel, science is full of instrumental value judgments. These seem akin to McMullin's "evaluating" and to Scriven's "value-performance claims": they state that a certain action is good in certain circumstances, that is, if a specified goal or value is to be obtained. Hempel and other positivists believed that categorical judgments of value, on the other hand, had no place in science. These categorical judgments are similar to McMullin's "valuing" and to Scriven's "real-value claims." They state that a certain goal is *prima facie* good, independent of particular circumstances.[22]

Although McMullin and Hempel distinguish between evaluation and valuing,[23] admittedly they come to different conclusions about the acceptability of *valuing* in science. Both accept evaluation or instrumental judgments of value in science, because they can be confirmed, because one can assess how well a particular means allows us to attain a certain end. We *evaluate,* for example, the fertility or simplicity of a particular theory. Hempel and other positivists rejected *valuing* in science, however, on the grounds that categorical judgments of value cannot be confirmed empirically. McMullin, Scriven, and other postpositivists accept valuing in science because they recognize that scientists attach different weights to different characteristic values—for example, simplicity, explanatory fertility. In the years of controversy surrounding Pauli's postulation of the neutrino, for example, scientists attached quite different weights to the value of internal coherence and, as a consequence, drew quite different conclusions about the plausibility of the neutrino hypothesis.

The great merit of McMullin's and Scriven's analyses is not only that they make precisely clear the sense in which value judgments are controversial in science, but also that they show why value judgments made in science do not necessarily throw one into a morass of relativism. Subsets of characteristic, epistemic values (values used in theory appraisal) are used in *evaluation* and *valuing* in science, and the skills of epistemic value judgment can be learned. On their accounts, one can learn how to use characteristic, epistemic values such as predictive accuracy, internal coherence, external consistency, unifying power, fertility, and simplicity, even though there is no algorithm for theory assessment.

USING A CASE STUDY TO EXAMINE VALUE JUDGMENTS IN SCIENCE

To show that we can rationally account for scientific progress and that we can sometimes rationally assess scientific values, I want both to argue that McMullin and Scriven are correct and to extend their analyses by providing examples of critieria for the types of value judgments for which they argue. Establishing both these points requires examining at least one case study. This case study will show the essential role played by both types of epistemic value judgments. Admittedly, however, as Scriven recognizes,[24] the dispute about value-free science focuses on *valuing,* not *evaluating* (on what Scriven would call "real-value claims" and not on "valued-performance claims"). This means that our case study, if it is to be interesting and valuable, must focus on the role of, and the justification for, valuing and real-value judgments in science.

Applied (rather than pure) science provides the best vehicle for illustrating my claims about a four-parameter model for justifying value judgments in science. Let's consider a case in hydrogeology to show that faulty inferences about methodological rules and cognitive values were responsible for erroneous scientific conclusions; that rational and impartial judges could probably agree on how to adjudicate these hydrogeological disputes over methods and goals; and that, if a rational justification procedure were used, in this case, then the faulty hydrogeological conclusions could have been avoided. Once the case study is completed, we can use it to draw some conclusions about instances in which rational assessment of cognitive values is possible in science.

A CASE IN HYDROGEOLOGY: MAXEY FLATS

The case study concerns attempts, twenty-five years ago, to ascertain subsurface migration rates for radionuclides at a proposed low-level radwaste storage facility. The science used to determine the alleged rates was fatally flawed, and geologists erroneously concluded that the site was suitable for shallow land burial of nuclear wastes. Although I am most interested in the epistemic value judgments that figured in the science underlying this conclusion, the case itself is an interesting one because it illustrates how flawed science, in this case, flawed hydrogeology, can lead to flawed policy.

Geologists gave the proposed site a "yes" vote, and the facility, in Maxey Flats, Kentucky, began accepting radwaste for burial (fifteen feet below the surface) in 1963. Because the site is underlain by virtually impermeable shales, Nuclear Engineering Company (NECO), now U.S. Ecology, originally said that it would take twenty-four thousand years for plutonium to migrate one-half inch at the site;[25] Several years later, NECO and California geologists at Emcon Associates testified before Congress that "the possibility of subsurface migration off-site is nonexistent" at Maxey Flats, and that on-site permeability would permit groundwater migration on site at a rate of "between 0.01 and 0.001 feet per year."[26] Even using the lowest predicted migration rates, it should have taken 1,161,000 years for plutonium in the trenches to move two miles off site. Yet, only ten years after opening the facility, plutonium and other radionuclides were discovered as far as two miles off site.[27]

The geological predictions at the proposed Kentucky site were wrong by *six orders of magnitude* and, as a consequence, the Maxey Flats facility has become the nation's "worst nuclear dump" (in terms of curies of off-site radionuclide migration). It contains more plutonium than any other commercial facility in the world, and nearby residents, largely uneducated farmers, have complained of an increasing number of unexplained lymph-node cancers, skin cancers, and mysterious deaths.[28] The 300-acre site has been predicted to become a "fountain of death for the next 75,000 years."[29]

Apart from its relevance to public-policy concerns, the Maxey Flats case is interesting because of its implications for scientific method and progress. How did judgments regarding methodological and cognitive values likely lead to a poor scientific conclusion and therefore to a poor siting decision. In particular, how can alternative judgments about weighting cognitive values, what McMullin calls *valuing* (which logical empiricists insist has no place in science), explain why the faulty geology occurred. If we can do so, and if we can show rational criteria for adjudicating these value conflicts, then we can establish not only that valuing takes place in science, but also that it does not necessarily interfere with scientific rationality and progress.

Before discussing the controversies over epistemic or cognitive values that arose in the hydrogeological investigations in the Maxey-Flats case, it is important to know how the scientists arrayed themselves on each side of the controversy. On the side of opening Maxey Flats and asserting its continuing suitability for radwaste storage, ten years after it was opened, were scientists such as Cohen from the University of Pittsburgh, Hajek from Auburn University, Moghissi from Georgia Tech, as well as a number of geologists from the Kentucky Department of Natural Resources, from the California geotechnical consulting firm, EMCON Associates, and from the site operator, NECO. On the side of those who proclaimed that Maxey Flats was unsuitable for radwaste storage and who maintained, after it was opened, that it should be closed, were a host of scientists from the U.S. Geological Survey (USGS), especially Zehner, and from the U.S. Environmental Protection Agency (EPA), especially Meyer.

Judgments about Methodological Values in the Maxey Flats Case

Many of the scientific disputes between the industry-EMCON-academic scientists and the USGS-EPA scientists were exactly of the sort that logical empiricists admit occur in science all the time. Several of these disagreements are significant enough to explain, in part, the erroneous scientific predictions made at Maxey Flats.

For example, the two groups disagreed both over whether exploratory drilling revealed that the shale was water bearing and over whether there was any perched groundwater. In Congressional testimony, the NECO side claimed that "exploratory drilling at depths of from 90 to 320 feet encountered no ground water in the disposal formation,"[30] and "no perched water at the contact of soil and the indurated Nancy [shale] formation."[31] All these claims were substantiated by the EMCON scientists.[32]

USGS geologists, on the other hand, contradicted the EMCON findings. They asserted that the water table appeared to range from between 2 to 6 feet below the surface at Maxey Flats,[33] to up to 15 to 20 feet below the surface,[34] and they pointed to various areas at the site at which they had found water.[35] The USGS project director, in a later report, claimed that there were at least two different saturated groundwater flow systems on site, one in the Ohio [to 185 feet] and Bedford Shales, and the other in the Nancy Shales.[36] However, at the early stages of the USGS investigation, they too believed that the site was dry, because their drill holes revealed no water.

This conflict over the facts, over whether the shales were water bearing and over whether there was any perched groundwater at Maxey Flats, can be resolved by an appeal to methodology, just as the logical-empiricist proponents of the hierarchical-triadic model affirm. Because groundwater movement in shale is extremely slow, and can occur only through fractures, if at all, and because the Maxey Flats hydrogeology is heterogenous in the extreme, the USGS scientists are correct in their methodological rule that such conditions demand both extremely long-term studies and extensive sampling, so as to include all areas of heterogeneity.[37] The USGS did the longest-term studies of the site, over a period of years, after the site was closed, and hence it is not surprising that they claimed to have found water in the drill holes and test wells. NECO and EMCON, on the other hand, did extremely short-term studies, with fewer drill holes and test wells, and so it is not surprising that their tests failed to yield evidence of groundwater in the disposal formation. Their drilling must not have intersected any fractures in the shale, because the fractures at Maxey Flats are anywhere from 1 to 40 feet apart, and run largely vertically.[38] In fact, NECO's siting decision appears to be based largely on a geological report done on the basis of only ten days of site monitoring.[39] Moreover, the same study based its conclusions about well recharge (after pumping) on only twenty-eight hours of observation.[40]

Because of site heterogeneity, short-term studies could easily miss drilling into a fractured area of the shale, and they could easily miss the perched groundwater. The USGS scientists found perched groundwater bounded on all sides by unsaturated zones, a fact that apparently led other scientists to assume that unsaturated conditions existed throughout the disposal formation.[41] USGS studies pointed out quite clearly that differences in water levels in the various test wells were a function of whether a fracture was exposed by the well, the number of fractures intersected, the vertical gradients, the presence of unsaturated zones at depth, and the length of the studies.[42] One central USGS report says quite clearly that "water tables may not be accurately located because wells are likely to intercept the saturated part of a fracture below the water table. . . . Water-level data collection could take years, because recovery must be measured over extended periods as small, incremental depths are drilled."[43]

If the USGS scientists are correct, and they appear to be, then factual controversy over the results of drilling and well-pumping tests could clearly explain how it was easy to miss the presence of water at Maxey Flats and how the water could transport radwastes off site. Moreover, it is also clear that this factual controversy over whether there is a perched water table and over whether the drill holes were dry can easily be solved by appeal to methodological rules requiring more tests and a longer period of testing. Were this methodological rule followed, and tests done over periods of years at Maxey Flats, then all scientists concerned should have arrived at roughly the same results as the USGS scientists. Hence, at least in this case, as the triadic hierarchical model of the logical empiricists affirms, factual disputes can often be resolved by appeal to methodological rules.

Let us turn now to a controversy over cognitive values, a type of conflict that logical empiricists allege ought not to occur in science. We shall discover both that such controversies are unavoidable in science and that, contrary to their claims, they can be rationally resolved.

Weighting Cognitive Values in the Maxey Flats Case

Obviously the EMCON-NECO scientists believed that the Maxey Flats site was hydrogeologically tight, whereas the USGS-EPA scientists believed that the site was not. A good example of how each group of scientists weighted cognitive values differently, in assessing their theories, occurred approximately ten years after the Maxey Flats site was opened. The occasion of their different weightings was the highly publicized claim of an EPA scientist that he had discovered plutonium two miles off site and that subsurface migration was responsible for his finding.[44] Neither side disputed the claim that, indeed, there was plutonium two miles off site and that it had come from the site. The EMCON-NECO scientists, however, held the view that the plutonium had come from surface runoff when site managers failed to clean up radwaste spills, and the USGS-EPA scientists held the view that the plutonium had gone off site through subsurface migration.

EMCON-NECO scientists supported the "surface" view by two theories: plutonium cannot migrate through the soil, especially at Maxey Flats; and soil permeability or hydraulic conductivity at Maxey Flats is too low to permit migration of anything out of the burial trenches. USGS-EPA scientists supported the "subsurface" view by two other theories: plutonium can migrate, contrary to the received opinion; and soil permeability or hydraulic conductivity at Maxey Flats is extremely low, but the chance occurrence of a fracture in the trenches could permit migration from the burial ground. Let's examine the respective theories about the plutonium claims to see how alternative groups of scientists weighted the two cognitive values differently.

When the Auburn, Georgia Tech, Pittsburgh, NECO, and EMCON scientists held to the theory that plutonium could not migrate in the soil, it appears that they did so because of adherence to the cognitive value of *external consistency,* consistency with previous studies confirming the immobility of plutonium in soil. When the USGS and EPA scientists held to the theory that plutonium could migrate in the soil, it appears that they did so because of adherence to the cognitive value of *internal coherence,* being unwilling to leave any coincidence unexplained by the theory that plutonium did not migrate. Let's examine the reasons for weighting cognitive values differently in assessing the plutonium theory.

In holding to the theory that plutonium could not migrate, the academic and industry scientists' main argument was that their theory was *externally consistent* with other findings. Among geologists and soil scientists, plutonium had the reputation of being absorbed on the soil within several centimeters of contact. The theory was that, the higher the pH, the more plutonium would be contained because of ion exchange and absorption by the surrounding soil and rock.[45] Reichert reported,[46] for example, that all of the radionuclides released to the seepage basins at the Savannah River Nuclear Plant, except plutonium, had been detected in the surrounding groundwater. Fennimore also concluded that plutonium does not move through soil,[47] even under a constant hydraulic head.[48]

EMCON and NECO scientists also cited studies by the best authors in the field, studies all coming to the conclusion that plutonium does not migrate deeper than 15 cm. below the surface in soils, unless the soils are disturbed;[49] and that downward movement of plutonium occurs only mechanically.[50] They also cited Hajek, who concluded that plutonium movement by contact diffusion would be less than 10 cm. after 240,000 years, and that a maximum of 0.1 percent of the plutonium would be leached by infiltrating groundwater.[51] Other scientists supported these same views about plutonium nonmigration, and Rhodes reported that soil sorption of plutonium was greater than 97 percent between pH 2 and 8;[52] Wilson and Cline[53] were unable to remove less than 1 percent of the plutonium in a soil column using standard leaching techniques.[54]

Given such studies, it was easy for the EMCON-NECO scientists to affirm that "the lateral migration of plutonium through the soil zone at Maxey Flats has not been established nor is there any conclusive evidence of plutonium migration through fractured systems in the subsurface geological formations.

The extra radiation found off-site was caused either by surface water runoff . . . or the result of cross contamination due to the use of improper procedures during sampling."[55] The surface contamination theory, as an alternative to subsurface migration of plutonium, was also reasonable because much of the plutonium, along with cobalt, strontium, and cesium, that was discovered off site was found in streambed sediment near the surface.[56] Moreover, the fact that the plutonium found in the test-well samples was associated with particulate matter (rather than being of a soluble nature, as it might be after having been in the trenches) also led the scientists to believe that it had arrived through surface runoff.[57] Even if one hypothesized that plutonium was transported, deep beneath the surface, by means of some organic compound, all sides agreed that the agent responsible for breaking down the organic plutonium complex is unknown.[58] Hence, the EMCON-NECO scientists reasoned that they had to side with extant scientific opinion, that subsurface plutonium transport was unlikely.

That the "subsurface migration" version of the plutonium theory was rejected on the grounds of external consistency is most apparent in a letter of one Georgia Tech scientist. He argues against lateral transport of plutonium in soil and says that "it is the view of the scientific community that plutonium remains essentially immobile in soil," and then he concludes his letter by citing the studies supporting this dominant position.[59] Also, the agency with perhaps the greatest expertise on movement of radionuclides, the Atomic Energy Commission, said explicitly that plutonium migration in soil or groundwater is unlikely.[60]

When the USGS and EPA scientists held to the theory that plutonium could migrate in the soil, however, they gave little weight to external consistency and to previous findings that plutonium could not migrate. They admitted that their theory affirming subsurface transport contradicted the scientific consensus of the day[61] instead, they said that this consensus theory was unable to account for a number of anomalies in the Maxey Flats plutonium discoveries. Hence it appears that they discounted surface migration of plutonium because of their adherence to the cognitive value of *internal coherence,* being unwilling to leave any coincidences unexplained by the theory that plutonium did not migrate through subsurface routes.

Perhaps the key unexplained fact, for which those who denied subsurface plutonium migration could not account, is that the USGS and EPA scientists found plutonium deeper in the soil than it should have been, if subsurface migration were not involved. According to the dominant theory, plutonium could not migrate more than a few centimeters in the soil; many tests, already cited, established that surface deposition of plutonium resulted in its remaining within several feet of the surface of the soil. Some EPA scientists on site at Maxey Flats also tended to substantiate this view, because their studies revealed that, as one drilled from 1.5 to 3.5 meters below the surface, the mean concentrations of all radionuclides decreased.[62] A key USGS scientist, however, discovered locations at which all site-related radionuclides, with the exception of plutonium, had higher median concentrations at a depth of 5 feet than at a depth of 10 feet; for plutonium, however the medium concentration was greater at 10 feet than at 5 feet.[63] Thus the "surface migration" theory could explain the deposition of all these other radionuclides, especially because it was known that median concentrations of the other radionuclides decreased with increasing distance of the sampling points from the evaporator; the evaporator was used to reduce amounts of contaminated water pumped out of the trenches.[64] However, the surface migration theory, whether postulating runoff or evaporator transport of radionuclides, could not explain the fact that plutonium concentrations were higher at greater depths.[65] Hence, according to the USGS scientists, there was no proof that the off-site plutonium contamination came from runoff alone or from cross contamination.[66]

As early as 1972, nine years after Maxey Flats was opened, the Kentucky Department of Human Resources found plutonium in the surface soil, soil cores 90 cm. deep, in monitoring wells, and in streams that drain the site.[67] All of these migrations of plutonium could be explained by "surface" theories, except for the plutonium in soil cores 90 cm. deep (since plutonium had been established not to move more deeply than 15 cm.).[68]

But if USGS and EPA scientists discounted the "surface" theory and alleged that plutonium migrated through subsurface routes, because of the anomaly that more plutonium had been found at greater drilling depths, how did they account for the alleged failure of ion-exchange theory according to which plutonium would not migrate? The answer is simple. They did not discredit the ion-exchange mechanisms, but simply said that, at Maxey Flats, the mechanisms were unable to work as expected because of some intervening variables. As one USGS geologist put it: at Maxey Flats "the ion exchange capacity along at least one pathway of migration has either been exceeded or . . . has been neutralized by some complexing agencies, such as organic material, or cleanup fluids, that were buried with the waste. If this, indeed, is the case, it would not be unreasonable to expect an increase in the concentration of the migration of radionuclides in the future."[69] Indeed scientists who monitored the site during its early days warned that fractures and fissures could allow the radioactive liquid to bypass the ion-exchange media more rapidly.[70]

Further, EPA and USGS scientists discounted earlier studies claiming that plutonium did not migrate by arguing that these studies were all concerned largely with the fate of plutonium deposited on soil surfaces and in semiarid regions. They criticized laboratory investigations of alleged plutonium migration for concentrating on column studies of plutonium leach rates and soil absorption capabilities of media occurring during the intergranular flow of a contaminated liquid through pulverized media. They admitted that virtually all researchers such as Reichert, Fennimore, Krey, Hardy, McClendon,

Pinder, Francis, Price, Hajek, Rhodes, and Wilson do agree that plutonium is retained in the soil column. The problem with such findings, however, according to EPA scientists, is that they fail to take account of what might happen to the plutonium in circumstances other than flow through a pulverized medium.[71]

A central EPA scientist, for example, noted that a number of agents and conditions, typically not considered in the extant research, can increase the mobility of plutonium buried in the ground. Some of these agents and factors cited by the EPA are anionic acetate complexes;[72] organic solvents, used during nuclear reprocessing operations;[73] acid soil;[74] chelating agents;[75] organic materials from plant root tissue;[76] and the presence of fractures or joints, because of the reduction in soil/rock surface area.[77]

Because there was evidence that at least the fractures and the requisite organic material were in contact with the trench water,[78] there were grounds for believing that the plutonium could have migrated through subsurface routes, although the agent responsible for the migration was not known. One Nuclear Regulatory Commission study concludes explicitly that microorganisms likely play a role in the movement of radionuclides because the radionuclides are not toxic to them.[79]

The controversy over the soundness of the theory that there is no subsurface migration of plutonium, as the preceding analysis reveals, is a controversy fraught with empirical underdetermination, both because most studies of plutonium migration were done on sandy soils and because no agent responsible for the alleged subsurface migration at Maxey Flats could be positively identified. The consensus of scientific opinion was on the side of the theory; hence, EMCO-NECO scientists decided, on grounds of the value of *external consistency,* to side with scientific consensus on the matter. Yet, the dominant scientific theory was unable to explain the anomaly that some of the plutonium concentrations increased with depth; hence USGS-EPA scientists decided, on grounds of the value of *internal consistency,* to disagree with scientific consensus on the matter. The disparate evaluations of the plutonium nonmigration theory apparently arose because different scientists attached different weights to different cognitive values. Hence, the plutonium nonmigration controversy appears to show, contrary to logical-empiricist belief, that scientists proceeding on rational grounds nevertheless engage in *valuing,* in weighting cognitive values differently.

This particular example also reveals that, if both sets of scientists had weighted the cognitive value of *internal consistency* as higher than *external consistency,* then they probably would have come to the scientific conclusion that is now believed to be correct—namely, that, given certain conditions, plutonium can migrate through subsurface routes. This means that our example has shown not only that scientists engage in *valuing* but that, more important, the way they engage in valuing can determine the success or failure of their scientific conclusions. But this raises the question of whether we can arrive at criteria for *valuing.* I believe that we can, and can thus guarantee the rationality and progress of science, although we cannot always specify these criteria ahead of time.

ADJUDICATING CONTROVERSIES OVER "VALUING" IN APPLIED SCIENCE

The main reason why scientists involved in the controversy over the plutonium migration theory weighted cognitive values differently is that both sides in the disputes had scientific information that was empirically under-determined and both sides had anomalies that their theories were unable to explain. Hence, in the plutonium migration dispute, the EMCON-NECO scientists relied heavily on the cognitive value of external consistency, while the USGS-EPA geologists appealed to the cognitive value of internal consistency. The positions of each group of scientists were clearly rational, in large part, because of the empirical underdetermination. As one Auburn scientist pointed out, the "scientific and technical merit" in the studies of both sides was questionable.[80] Because both sides had anomalies, and because there were no "smoking guns," both groups of scientists appear to have been behaving rationally in adhering to their positions.

But if both sides were behaving rationally in weighting cognitive values differently, then how might they have been able to arrive at the same weighting for the cognitive values that they used in assessing their theories? The logical-empiricist proponents of the hierarchical model would claim that there is no criterion by virtue of which both sides might arrive at the same weighting of cognitive values. Their claim is premature, however, in part because, as Laudan points out,[81] scientists are sometimes able to choose among cognitive values, as when a particular value is unrealizable, and therefore incapable of being weighted heavily in a given situation. One need only think of numerous situations in ecology, for example, to realize that often the cognitive value of predictive accuracy cannot be realized and therefore has to give way to other values.

It is also pessimistic and premature, especially in cases of applied science, to claim that there is no criterion by virtue of which both sides in a scientific dispute might arrive at the same weighting of cognitive values. Applied science, simply by virtue of the fact that it is applied, always has an end or goal by means of which alternative cognitive values might be assessed. Such ends or goals might be determining whether a toxic drug was present in lethal quantities, or determining whether a specific building structure could withstand an earthquake of a particular magnitude.

In the case of the applied science hydrogeology, used to investigate theories about plutonium migration and effects of permeability at the Maxey-Flats site, the practical end or goal was quite clear; the scientific conclusions were the *means* to the *end* of protecting the public from off-site radiation. All discussions of the relevant geology were predicated on this goal, protecting "public health and welfare," especially because the biological hazard from radioactive waste, due to the effects of ionizing radiation, could last in perpetuity. Half-lives of some of the isotopes in question are as great as 15,900,000 years (I-129), 2,300,000 years (Cs-135), and 23,420,000 years (U-236), not to mention the 24,390-year half-life of plutonium 239.[82] Because of the long half-lives, any mistake in estimating groundwater-transport rates of such radionuclides could have disastrous consequences for centuries. Avoiding these disastrous consequences was perhaps the key end or ethical value of the Maxey Flats' application of hydrogeology.

Another way of formulating this end of the applied science would be to say that the hydrogeological conclusions were means to the end of *conservatism* regarding the imposition of public (and therefore involuntary and uncompensated) risk. But if so, then cognitive values such as simplicity and external coherence were (in part) assessed and weighted on the basis of how well they served as means to this conservative ethical end.

There is some evidence that this ethical end of risk conservatism did serve as a basis for evaluating the cognitive goals weighted most heavily by the USGS and EPA scientists. One key EPA scientist, in the central Maxey Flats document on radionuclide migration, specifically stated: "If 100 percent retention of a waste for its hazardous lifetime is the goal of shallow land disposal, continued burial of Pu (and other radionuclides) in humid climates [such as occurred at Maxey Flats] using present waste forms, containers, and trench construction methods will not achieve this goal."[83] Moreover, at the beginning of a major report on the site, the USGS project director said specifically that he expected the USGS findings to be "an aid in interpreting monitoring data gathered by State and Federal agencies responsible for regulation of the Maxey Flats site, and protection of the environment around the site."[84] In another important hydrogeological study of the Maxey-Flats site, the USGS project director mentioned that the study was undertaken at the request of the U.S. government, to evaluate the radwaste site with respect to long-term considerations of stability and safety.[85] Statements such as these show that the EPA and USGS scientists were clearly concerned about public safety and the public-policy impact of their scientific conclusions; hence, it is likely that they viewed their science as a means to the end of public health and safety.

But if USGS and EPA scientists viewed their science as a means to the end of public health and safety, then whenever they were in a situation of empirical underdetermination, it is probable that they would choose to err on the side of risk conservatism, so as to be assured of protecting the public. For example, as is clear from the preceding discussion, both the EMCON-NECO scientists and the USGS-EPA scientists admitted that the radwaste trenches were dug in the Nancy Shale and that the Nancy Shale was "very impermeable." Only as little as a foot below this shale, however, was the sandstone marker bed, "one of the most permeable units at the site"; it occurs "directly beneath the trenches."[86] The USGS scientists explicitly said that they were worried about this sandstone-marker bed, and that "the possibility of radionuclide movement in the sandstone marker bed is of major concern."[87] No one knew for sure whether the EMCON-NECO scientists were correct in arguing for the impossibility of off-site migration, or whether the USGS-EPA scientists were correct in using internal consistency to argue that off-site migration was possible. If the EMCON-NECO scientists were wrong, however, their mistake could be disastrous because of the close proximity of the sandstone beneath the trenches. Hence, in part because of their admitted concern about the nearby sandstone, it appears that the USGS scientists chose to err, if at all, through risk conservatism in dealing with a scientifically controversial question like waste migration.[88]

If the applied science at Maxey Flats was, and ought to be, ultimately assessed in terms of how well it enabled regulators to protect the public, then, at least in such applied cases, there might be a fourth, or ethical, means of adjudicating scientific controversy. Just as factual disputes often are adjudicated by appeal to methodology, and as methodological disputes frequently are evaluated by means of cognitive goals, so also it might be that some disputes over cognitive values ought to be adjudicated by means of ethical values, at least in cases of controversies concerning applied science. In other words, disputes over weighting cognitive values might be adjudicated in terms of how well recognition of each specific cognitive value contributes to the ethical end or value for which the scientific study is undertaken. In the Maxey Flats case, this ethical value was protecting the public. In fact, in virtually every instance of *applied science* put at the service of public policy, one could argue (although I shall not take the time to do so here) that the most important ethical value, in terms of which to assess cognitive scientific values, is protecting public welfare. As Cicero put it: "Salus populi suprema est lex." That is, "the people's health and safety is the highest law."[89]

McMullin points out that ethical values [like protecting the public] involve not epistemic assessment, but calculation of *utilities* of some sort, because scientific theory is being applied to practical ends, and the theoretical alternatives carry with them outcomes of different value to the agents concerned. Such utilities are irrelevant to theoretical science, as theoretical science, notes McMullin, and they cannot be cited as a reason for holding that science itself is laden with ethical values.[90]

Despite McMullin's warnings, there might be reasons for thinking that science as science, pure science, does involve scientists in using ethical values to assess alternative cognitive values like explanatory fertility or simplicity. I can think of at least one such ethical value, the duty to be open-minded in assessing novel, controversial, or innovative scientific results. Such a duty is obviously ethical, because it is closely akin to the duty to promote fair play, or equal consideration of all scientists' interests. It also is an ethical value because the consequences of recognizing (or not recognizing) it have effects on the welfare of others—for example, on the proponents of the innovative scientific theories. In other words, the duty to be open to novel scientific theories helps to increase the probability that novel theories will be recognized. This recognition, in turn, obviously has value to the proponents of the novel theories, just as recognition of particular hydrogeological theories obviously has value both to their proponents and (in applied cases, like Maxey Flats) to the public.

The *ethical* value, the duty to be open to innovative theories, also contributes to science as a purely *epistemic* activity. This is clear if one considers what would happen were there no duty to be open-minded; science would likely not progress as rapidly as it might, dogmatism and ideology might take the place of empiricism and testing, and Russian Lysenkoists might become more the rule than the exception.

J. S. Haldane tells an interesting tale of what happened in one case of failure to be open to innovative results. The referees of the Royal Society rejected some groundbreaking papers of the nineteenth-century English chemist, J. J. Waterson. One of the referee's comments was that "the paper is nothing but of nonsense." Haldane wrote:

> It is probable that, in the long and honorable history of the Royal Society, no mistake more disastrous in its actual consequences for the progress of science and the reputation of British science than the rejection of Waterson's papers was ever made. The papers were foundation stones of a new branch of scientific knowledge, molecular physics, as Waterson called it, or physical chemistry and thermodynamics as it is now called. There is every reason for believing that, had the papers been published, physical chemistry and thermodynamics would have developed mainly in this country [i.e., England], and along much simpler, more correct, and more intelligible lines than those of their actual development.[91]

As the remarks of Haldane illustrate, openness to innovative theories is clearly an epistemic value. But if the duty to be open to innovative science is both an epistemic as well as an ethical value, as was argued previously, then science as science, pure science, may involve ethical, as well as epistemic values. And if pure science involves ethical values, at least in situations where the ethical value of openness to innovative theories is a way of resolving conflicts among cognitive values, then we may need a new model for scientific rationality. We may continue to adjudicate factual disputes by means of methodological rules, and to evaluate methodological judgments by means of cognitive values. But if the suggestions in this essay are correct then, even in pure science, we may sometimes be able to adjudicate disputes about cognitive goals in terms of a fourth type of judgment, those about ethical values.

For example, when the EMCON-NECO scientists were at odds with the USGS-EPA scientists over whether subsurface plutonium migration was possible, the USGS-EPA scientists were clearly the ones with the innovative, unpopular theory. External consistency was on the side of the EMCON-NECO scientists. Yet if this controversy over internal versus external consistency had been adjudicated in terms of the ethical value of openness to innovative theories, then it is likely that hydrogeology would have come sooner to the theory now believed to be most correct. In other words, the Maxey Flats dispute over cognitive values could have been adjudicated in terms of a particular ethical value (openness to innovative theories). Moreover there is reason to believe that it ought to have been adjudicated in terms of this ethical value, because such openness would promote theoretical pluralism. Especially in a situation rampant with anomalies, theoretical pluralism appears to be desirable.

OBJECTIONS TO THE NEW MODEL OF SCIENTIFIC JUSTIFICATION

In response to the quadrilateral model of scientific justification, a number of objections can be made. One of the most important is that use of ethical values is accidental, not essential, to the practice of science, or that the value judgment that I have used in my last example is "accidentally ethical."

This first objection very likely arises from the long and important tradition of attempting to keep values, especially emotive or bias values, out of science, so as to safeguard the ideal of scientific objectivity. Despite the importance of this ideal, however, it is obvious that some values, for example, cognitive or espistemic ones (as discussed earlier in this essay), are essential to the practice of science. Moreover, in at least one sense, it also appears that making judgments about ethical values is essential to science. The reasoning is as follows. Science cannot be done without scientists. Yet scientists as scientists continually must make judgments about issues, such as theory acceptance and weighting available evidence. But when scientists as scientists must make judgments about theory acceptance and weighting available evidence, they must make decisions about how open they ought to be to innovative theories and anomalous data. But decisions about openness in these matters are in part ethical decisions, primarily because they have consequences for the welfare of the proponents of the innovative theories (and perhaps also for the welfare of the public); they presuppose particular notions of fair play and equal opportunity in the "game" of science; and they presuppose an implicit notion of procedural justice, an implicit notion of a correct method for arriving at any distribution—for example, of research monies or of professional (nonmonetary) support from ones's peers. For all these reasons ethical judgments about openness to innovative scientific theories or data are essentially, though only in part, ethical judgments.

A second possible objection is that the fourth type of scientific justification, that achieved by means of ethical values and goals, is not useful because, if there is to be consensus about ethical values, the values in question must be fairly obvious and noncontroversial. But if they are fairly obvious and noncontroversial, so the objection goes, then they are not useful in resolving disputes over cognitive goals or values. If, however, the ethical judgments are controversial and not obvious, goes the objection, then they would require a criterion for assessing their correctness. But there is no such criterion.[92]

In responding to this second objection, it is important to point out that establishing that ethical values can sometimes be used as criteria to settle disputes over cognitive values does not wholly deliver science from the charges that it is not progressive or not amenable to rational adjudication of disputes. This is because the fourth or ethical type of dispute, as the previous objection recognizes, would itself need a criterion for resolution. Adding a fourth parameter (ethical goals or values) to the model of rational justification in science thus keeps the irrationalists at bay, if at all, by putting one more stumbling block, one more means of adjudication, in their path. It certainly does not end all controversy in science. It does, however, make the nature of some scientific disputes partially clearer than they were under the logical empiricists' hierarchical model and Laudan's triadic model. The nature of the disputes is clearer because we have identified the ethical locus of many controversies. Hence, at least in this sense, the new model is useful both because it helps clarify disputes in applied science and because it points up the relevance of ethical judgments, a fact often not recognized.

Moreover, to claim that there is no specific criterion for adjudication of disputes about ethical values or goals in science is not as troublesome as it sounds. At least in the case of applied science, the ethical goals of the research are typically set by the agencies funding the relevant studies. For the research treated in this essay, the USGS and the EPA explicitly formulated this goal as protecting public health and safety. Admittedly, in the case of pure science (if there is such a thing), the ethical goals and values would need to be established by analysis. Although such an analysis could be problematic, it would not be impossible, both because ethical theory is well developed and not dependent upon new discoveries, as is resolution of some conflicts over largely methodological issues in science, and because codes of professional ethics for scientists often function as guides to behavior.

A third objection, one that attacks not only my model of scientific justification but all such models, including those of the logical empiricists and Laudan, is that it is difficult, if not impossible, to know ahead of time which methodological rules or cognitive goals ought to be employed in a particular situation. In other words, because there is no specific algorithm for doing correct science, one cannot tell, ahead of time, which methodological rule or cognitive goal will help deliver the correct conclusion in science.

The main problem with this third objection, as Scheffler and others have pointed out, is that it presupposes that, unless one can specify all methodological rules and cognitive goals ahead of time, then the rules and goals are not very useful in adjudicating conflicts in science. This presupposition is false, however, because scientific rationality does not require that we be able to predict specific methodological rules or cognitive goals ahead of time, but merely that we know ahead of time the *general* (not the specific) methodological rules. Or, as Scheffler puts it, rationality is tied to "a commitment to general rules capable of running against one's own wishes in any particular case"; rationality means that one is committed to using general rules in science, as a basis for decisions, instead of merely doing what one wants, for whatever reason.[93]

One such *general* methodological rule, defended by a great many philosophers of science, including Hempel and McMullin, is to choose the theory with the greatest explanatory power, as tested by prediction. With respect to *particular* methodological rules, scientific rationality requires only that we be able to make reasonable choices in the specific situation, as Scriven and others have pointed out. Scientific rationality requires neither that we have infallible methodological rules, nor that we are able to formulate the specific rules ahead of time.[94]

Related to the notion of scientific rationality is a fourth objection, that this essay appears to presuppose a notion of "objective" resolution of scientific controversies, by means of value judgments. Yet, maintains the objector, the essay defends no value-free notion of objectivity and no set of absolute rules.

The main flaw in this objection is that it appears to presuppose that value judgments cannot be objective in any sense, that "objective" must mean "value-free," and that objectivity must be guaranteed by an appeal to some "absolute" rule or rules. All three presuppositions are false.

It is false to assume that value judgments cannot be objective in any sense; for one thing, to deny that value judgments can be objective would be to claim that one value judgment is as good as another. Yet it is obvious that, given two value judgments, there are many cases in which we can say that one is better than the other, even if we have difficulty specifying the rules on which we base our decision. For example, given the value judgments, (A) "Torturing innocent children for no good reason is a good way to spend a Sunday afternoon," and (B) "Torturing innocent children for no good reason is wrong," obviously (B) is a better or more correct moral judgment. But if so, then we must be able to rank at least some value judgments on a scale of better to worse. But if so, then at least some value judgments must be objective in some sense, regardless of whether we are able to specify, ahead of time, what our criteria for those judgments are.

Moreover, there are a number of value judgments that virtually everyone would admit are both true and objective. For example, suppose I say, "Einstein was one of the most brilliant physicists of this century." Obviously this is a value judgment. Equally obviously, this value judgment is objective. The most objective thing to say about a brilliant physicist is that he is a brilliant physicist. If a physicist is brilliant, and we describe his abilities in purely neutral terms, then we are not serving the ideal of objectivity. We are confusing objectivity with neutrality. Sometimes the most objective value judgments are anything but neutral, because the situations they describe or assess are often not neutral.

It is also wrong to assume that "objective" means "absolute" or "value-free." If Hanson, Kuhn, Polanyi, and others are correct, no concepts and no propositions are value-free or absolute in the sense of presupposing no theoretical assumptions. And if not, then it is unreasonable to define "objective" as "absolute" or "value-free," because we would be unable to describe any judgment as objective.

A more reasonable way to define objectivity, given a world of propositions, none of which are value-free, is as Popper and Scheffler have done. Perhaps Scheffler puts it best: "Objectivity requires simply the possibility of intelligible debate over rival paradigms.":[95] For Popper, a statement in science can be said to be "objective" in much the same sense as Scheffler affirms: if it is capable of being subjected to the criticism and the debate of the scientific community.[96] Scheffler and Popper do not demand that objective statements be absolute and value-free. They don't presuppose infallibility, but only what reasonable people, behaving reasonably, do. Scheffler writes: "I have, at no time, any guarantees that my system will stand the test of the future, but the continual task of present evaluation is the only task it is possible for me to undertake. Science, generally, prospers not through seeking impossible guarantees, but through striving to systematize credibly a continuously expanding experience.[97]

Although I do not have time to defend it here, on my account, *scientific rationality* is guaranteed by *individual* scientists, pursuing the general goal of choosing theories on the basis of explanatory power, tested by prediction. *Scientific objectivity,* however, can only be guaranteed by the *community* of scientists, because the community as a whole must provide the debate and criticism necessary to evaluate a theory choice, methodological rule, or goal in science. Admittedly, this sense of "scientific objectivity" is not the only one, and perhaps it is not even the most important one. My claim is merely that, in at least one interesting sense of "objectivity," scientists guard against prejudice and bias, and therefore guarantee objectivity (but not infallibility), by means of open, rational debate. In appealing to the fact that we make objective value judgments every day, that we know how to rank many value judgments as better or worse, and that objectivity requires the possibility of rational debate and criticism, I am presupposing, along with many epistemological naturalists, that evaluating alternatives in science and in ethics is not radically different from other kinds of decision making. The situation itself, along with common sense, often dictates correct decisions, even though we cannot produce, ahead of time, *specific* algorithms for guaranteeing the correctness of those decisions.[98]

A fifth objection to my model of scientific justification is that subscribing to the ethical value of being conservative regarding the imposition of public risk, as in the Maxey Flats case study, often could lead to false scientific conclusions. Following the ethical goal of being conservative regarding risk imposition, for example, the US Nuclear Regulatory Commission in 1978 shut down General Electric's Vallecitos (California) Test Reactor for six years because of an alleged geological fault nearby. The shutdown was accomplished at considerable expense to GE and subsequent loss of its medical isotope business. After the six-year period, the fault was shown not to be a problem; hence risk conservatism dictated the wrong scientific position.[99] More generally, continues the objection, following a position of risk conservatism could lead to lack of progress in science, costly technological delays, and erroneous conclusions.

As this objection correctly recognizes, adopting the ethical goal of being conservative in risk imposition can lead to erroneous scientific conclusions. The possibility of erroneous scientific conclusions, however, no more invalidates the employment of this ethical goal than the possibility of erroneous jury decisions invalidates the use of trial by one's peers. In both cases, the relevant issue is finding the most desirable model or option, given a situation in which no scientific or juridical options are perfect. This means that, in order to defend the ethical goal of risk conservatism, it is not necessary to prove that adopting the goal would never result in faulty scientific conclusions, but only that using this goal is better than not using it. Because the types of risk under consideration in areas of applied science are typically public, uncompensated, and involuntarily imposed on potential victims without their specific consent, it is not difficult to prove that, all things being equal, adopting a position of risk conservatism is more desirable than not adopting it.

Risk conservatism in cases of public risk—for example, opening a hazardous waste site or building a liquefied natural gas facility or a nuclear plant—often is required on grounds of informed consent. If acceptance of facilities like these is not subject to the free, informed consent of all the individuals put at risk by them, then ethics typically dictates, at least, that the risk be minimized, if not prohibited. In other words, because such consent is never (and probably can never be) obtained from all potential victims of such facilities, risk conservatism often is required.[100]

Moreover, in the case of catastrophic accidents associated with most public risks, for example, in case of a nuclear core melt, industry liability typically is limited by law so as to protect the corporation from bankruptcy.

This means that, in precisely the cases of public risk most often associated with applied science (science used at the service of public policy), potential victims arguably do not have due-process rights. But if certain types of risk imposition might jeopardize due-process rights, and if due-process rights are required both by civil convention and by morality, then there are strong ethical and due-process grounds for adopting a conservative position regarding the imposition of public risk, such as that relevant to applied science.[101]

Moreover, at least one other factor might mitigate possible negative consequences of adopting a goal of risk conservatism. The ethical goal of adopting a conservative position regarding the imposition of public risk might be counterbalanced by another ethical goal. More specifically, although risk conservatism might make scientists reluctant to advise implementation of new science and technology, another ethical goal (openness to innovative results) might operate so as to encourage new approaches to science, its applications, and to technology. Hence the two ethical goals of conservatism and openness could operate to balance each other.

Related to the "conservatism" problem is another objection: How might one qualify or limit the ethical values proposed in the essay, among them, openness to innovative theories and risk conservatism? Obviously, goes the objection, these values are not absolute in the sense that they are applicable in every situation and overrule every contrary consideration.

Although the two ethical rules I have proposed need to be interpreted in every situation in which they are applied, it is impossible to specify precisely all the rules of interpretation that might be useful, for example, when a risk is so insignificant that one does not have to be conservative regarding the desirability of its imposition. At least two points might be helpful, however, in providing an initial understanding of how to apply the rules.

The first point is that the rules are not absolute. Because they are not, they are best thought of as *prima facie rules,* rules that dictate a course of action, all other things being equal. Moreover, as *prima facie rules,* one of their main assets is that they place the burden on proof on the person who wishes to deviate from adherence to them.

Second, although the interpretational context for both rules needs to be developed more fully, John Rawls provided a good example of the sorts of conditions that would need to be met, in a particular situation, in order to justify a conservative stance regarding the imposition of public risk. He argues that such a stance is reasonable whenever the probabilities associated with various outcomes are uncertain; the benefits to be obtained either are not great or are not desired; and the situation involves great risks.[102]

Note, however, that for my model of scientific justification to be plausible, I need not specify all the relevant conditions under which particular ethical goals or rules ought apply. Rather, I need show only that reasonable people, in the arguably typical case study I have described, would be likely to reason as I have suggested and for some of the reasons that I have suggested.

CONCLUSIONS

In this essay I have attempted to argue that there are criteria for rational analysis of cognitive values in science. As the case study in hydrogeology shows, in *applied science* these criteria are often practical or ethical in nature (e.g., protecting the public) and are specified by the purpose for which the scientific research is being done. In *pure science,* at least one such criterion concerns openness to innovative theories. If I am correct in believing that there are rational criteria for assessing cognitive values in science, then it is plausible to believe that there may be criteria both for rationality in science and for progress in science, at least in the sense that such criteria often can tell us what we ought *not* to do.

Whether or not my analysis has been correct in arguing that ethical values have a clear role both in applied science and in certain cases of pure science (namely, where openness to innovative scientific theories is relevant to assessing cognitive values), it suggests several possible lines of fruitful inquiry.

First, philosophers of science might want to determine whether pure science exhibits any other values (like openness to innovative scientific theories) that are either ethical or both ethical and epistemic values. Second, they might want to investigate a number of scientific controversies over cognitive values in order to determine the conditions under which the ethical value discussed here (openness to innovative scientific theories) might function as a criterion for adjudicating the conflicts.

If both these possible inquiries appear promising, then there might be further evidence for my claims that judgments about ethical values are essential even to some instances of pure science and to most cases of applied science; that the model for scientific justification might sometimes include a fourth, or ethical, type of judgment; and that there are often rational criteria for assessing conflicts over cognitive values in science.

If my analysis is correct, then the hierarchical model of the logical empiricists errs, as Laudan argued, in assuming that there are *levels* of justification. Likewise, Laudan's triadic model errs through incompleteness, because it fails to account for justification by means of ethical goals and values, especially in applied science. If my analysis is correct, then, at least in some clearly defined cases, scientific disputes can and ought to be adjudicated by means of ethical values. But if so, then we may have to rethink not only our conception of science, but also our theories about the relationship between ethics and pure science.

NOTES

I am grateful to Steven Goldman for numerous helpful comments that served to strengthen this paper; whatever errors remain are, of course, my responsibility.

1. See Larry Laudan, *Science and Values* (Berkeley: University of California Press, 1984), p. 47; hereafter cited as Laudan, SV.

2. Ibid., p. 5.

3. Thomas Kuhn, *The Structure of Scientific Revolutions* (Chicago: University of Chicago Press, 1962), hereafter cited as Kuhn, *Revolutions;* and Kuhn, *The Essential Tension* (Chicago: University of Chicago Press, 1977).

4. Paul Feyerabend, *Against Method* (New York: Schocken 1978), hereafter cited as Feyerabend, AM.

5. Ian Mitroff, *The Subjective Side of Science* (New York: Elsevier, 1974).

6. Laudan, SV, pp. 47–49.

7. Karl Popper, *Logic of Scientific Discovery* (New York: Basic Books, 1959), hereafter cited as Popper, LSD; and Hans Reichenbach, *Experience and Prediction* (Chicago: University of Chicago Press, 1938).

8. Laudan, SV, p. 57; and E. McMullin, "Values in Science," in *PSA 1982*, vol. 2, ed. Peter Asquith (East Lansing, Mich.: Philosophy of Science Association, 1982), hereafter cited as McMullin, VIS.

9. See Feyerabend, AM, pp. 23, 28; Kuhn, *Revolutions,* pp. 3–5, 57, 62–64; and Laudan, SV, p. 70.

10. Israel Scheffler, *Science and Subjectivity* (New York: Bobbs-Merrill, 1967), pp. 1–3, 5, 18–19, hereafter cited as Scheffler, SS. Thanks to Steve Goldman for pointing out this Schefflerian insight.

11. Peter Caws, *Science and the Theory of Value* (New York: Random House, 1967), p. 59, hereafter cited as Caws, STV.

12. Ibid., p. 61.

13. M. Scriven, "The Exact Role of Value Judgments in Science," in *Proceedings of the 1972 Biennial Meeting of the Philosophy of Science Association,* ed. R. S. Cohen and K. Schaffner (Dordrecht: Reidel, 1974), pp. 219–47, hereafter cited as Scriven, VJ; and Caws, STV, pp. 68–69.

14. Caws, STV, p. 54.

15. Ibid., p. 55.

16. Ibid., p. 61.

17. Ibid., p. 62.

18. Helen Longino, "Beyond Bad Science: Skeptical Reflections on the Value-Freedom of Scientific Inquiry," unpublished essay, March 1982, done with the assistance of National Science Foundation Grant OSS 8018095.

19. See R. Carnap, "The Elimination of Metaphysics through Logical Analysis of Language," in *Logical Positivism,* ed. A. J. Ayer (Glencoe, Ill.: Free Press, 1959), pp. 60–81.

20. See R. Rudner, "The Scientist *qua* Scientist Makes Value Judgments," *Philosophy of Science* 20 (1953): 1–6.

21. McMullin, VIS.

22. Carl Hempel, "Science and Human Values," in *Readings in the Philosophy of Science,* ed. E. D. Klemke, R. Hollinger, and A. Kline (Buffalo, N.Y.: Prometheus Books, 1980) pp. 254–68, hereafter cited as Klemke, Hollinger, and Kline, IRPS; and Scriven, VJ.

23. Following E. Nagel, *The Structure of Science,* (New York: Harcourt Brace, 1961), p. 492.

24. Scriven, VJ.

25. U.S. Geological Survey, vertical file, "Maxey Flats: Publicity" (Louisville, Ky.: Water Resources Division of the Interior, 1962), this vertical file hereafter cited as USGS. (The Louisville office of the USGS is responsible for monitoring the Maxey Flats radioactive facility.) See A. Weiss and P. Columbo, *Evaluation of Isotope*

Migration—Land Burial, NUREG/CR-1289 BNL-NUREG-51143 (Washington D.C.: U.S. Nuclear Regulatory Commission, 1980), hereafter cited as Weiss and Columbo, EIM. At this time, transuranics in small quantities were included among low-level wastes.

26. J. Neel, "Testimony," in U.S. Congress, *Low-Level Radioactive Waste Disposal,* Hearings before a subcommittee of the Committee on Government Operations, House of Representative, 94th Congress, Second Session, February 23, March 12, and April 6, 1976, Washington, D.C., U.S. Government Printing Office, 1976), p. 258; hereafter cited as: US Congress and Neel, 1976.

27. G. Meyer, "Maxey Flats Radioactive Waste Burial Site: Status Report," unpublished report (Washington, D.C.: Advanced Science and Technology Branch, U.S. Environmental Protection Agency, 19 February 1975), p. 9, hereafter cited as Meyer, SR, 1975.

28. W. F. Naedele, "Nuclear Grave is Haunting Kentucky," *Philadelphia Bulletin,* 17 May 1979, pp. 1–3, in USGS, 1962.

29. Frank Browning, "The Nuclear Wasteland," *New Times* 7 (July 1976): 43.

30. Neel, 1976, p. 258.

31. B. F. Hajek, "Plutonium Mobility on Soils," BNWL-cc-925 (Washington, D.C.: U.S. Atomic Energy Commission, 1966), p. 272, hereafter cited as Hajek, PMS, 1966.

32. Jack McCollough and EMCON Associates, *Geological Investigation and Waste Management Studies, Nuclear Waste Burial Site, Fleming County, Kentucky,* Project 108-52, unpublished report, 6 February 1975, available from EMCON, 326 Commercial Street, San Jose, Ca.

33. H. Hopkins, "Ground-Water Conditions at Maxey Flats, Fleming County, Kentucky, unpublished report, U.S. Geological Survey, Water Resources Division, 1962, p. 2, in USGS.

34. H. H. Zehner, *Hydrogeologic Investigation of the Maxey Flats Radioactive Waste Burial Site, Fleming County, Kentucky,* USGS Open-File Report, Louisville, draft, 1981, pp. 96–98, hereafter cited as Zehner, HI, 1981.

35. H. H. Zehner, hydrologist, USGS, "Letter to C. M. Hardin, Manager, Radiation Control Branch, Kentucky, Department for Human Resources, February 6, 1975," p. 3, in USGS.

36. D. W. Polluck and H. H. Zehner, "A Conceptual Analysis of the Ground-Water Flow System at the Maxey Flats Radioactive Waste Burial Site, Fleming County, Kentucky," USGS Open-File Report, Louisville, Ky., 1981, pp. 1–5, hereafter cited as Polluck and Zehner, CA, 1981. (Published in C. Little and L. Stratton, eds., *Modeling and Low-Level Waste Management,* ORO-821, available from National Technical Information Service, U.S. Department of Commerce, Springfield, Va., 22161.)

37. G. D. DeBuchananne, Chief, Office of Radiohydrology, Water Resources Division, USGS, "Statement," in U.S. Congress, 1976, p. 135, hereafter cited as DeBuchananne, "Statement."

38. Polluck and Zehner, CA, 1981, p. 3.

39. I. Walker, *Geological and Hydrologic Evaluation of a Proposed Site for Burial of Solid Radioactive Wastes Northwest of Morehead, Fleming County, Kentucky,* 12 September 1962, unpublished report, p. 3, hereafter cited as Walker, GHE, 1962. This is on file in the Louisville, Ky., office of the USGS, in USGS.

40. Ibid., p. 15.

41. Polluck and Zehner, CA, 1981, p. 4.

42. Zehner, HI, 1981, pp. 110–12.

43. Ibid., p. 179.

44. Meyer, SR, 1975.

45. Kentucky Science and Technology Commission, "Technical Review of the Maxey Flats Radioactive Waste Burial Site," unpublished report, Kentucky Department of Human Resources, Frankfort, Ky., 1972, pp. 7–8, hereafter cited as KSTC, 1972.

46. S. O. Reichert, "Radionuclides in Groundwater at the Savannah River Plant Waste Disposal Facilities," *Journal of Geophysical Research* 67 (1962): 4363ff.

47. J. W. Fennimore, "Land Burial of Solid Radioactive Waste During a 10-Year Period," *Health Physics* 10 (1964): 239ff.

48. G. Meyer, *Preliminary Data on the Occurrence of Transuranium Nuclides in the Environment at the Radioactive Waste Burial Site, Maxey Flats, Kentucky* (Washington D.C.: U.S. Environmental Protection Agency, Office of Radioactive Programs, February 1976), p. 44, hereafter cited as Meyer, PD, 1976.

49. P. Krey and E. Hardy, "Plutonium in Soil Around the Rocky Flats Plant," *Health Physics* 28 (1975): 347ff., hereafter cited as McClendon SRP; and H. L. Pinder, et al., "A Field Study of Certain Plutonium Contents of Old Field Vegetation and Soil Under Humid Climatic Conditions," unpublished proceedings, Radiology Symposium, Corvallis, Oreg., 1975; hereafter cited as Pinder, "Field Study."

50. C. E. Francis, "Plutonium Mobility in Soil and Uptake in Plants: A Review," *Journal of Environmental Quality* 2 (1973): 62ff., hereafter cited as Price, PM, 1973.

51. Hajek, PMS, 1966.

52. D. Rhodes, "Absorption of Plutonium by Soil," *Soil Science* 84 (1967): 465ff., hereafter cited as Rhodes, APS.

53. D. O. Wilson and J. F. Cline, "Removal of Plutonium-239, Tungsten-185, and Lead 210 from Soils," *Nature* 209 (1966): 941ff.

54. Meyer, PD, 1976, p. 45.

55. Neel, 1976, p. 257.

56. Zehner, HI, 1981, p. 58.

57. R. L. Blanchard, D. M. Montgomer, H. E. Kolde, and G. L. Gels, "Supplementary Radiological Measurements at the Maxey Flats Radioactive Waste Burial Site—1976 to 1977," EPA-520/5-78-011 (Montgomery, Ala.: Office of Radiation Programs, U.S. Environmental Protection Agency, 1978), p. 29, hereafter cited as Blanchard, et al., SRM.

58. Blanchard et al., SRM, p. 29.

59. Al Moghissi, "Letter to H. Holton, NECO, 17 Feburary 1976," U.S. Congress, 1976, pp. 269.

60. Meyer, PD, 1976, p. 2.

61. Ibid., pp. 44–45.

62. Blanchard et al. SRM. p. 29.

63. Zehner, HI, 1981, p. 104; see also Meyer, PD, 1976, and Blanchard, et al., SRM, pp. 1–5.

64. Zehner, HI, 1981, p. 147.

65. Ibid., p. 104.

66. H. H. Zehner, hydrologist, USGS, "Notes in the Margin of U.S. Congress, 1976," available from Water Resources Division, U.S. Division of the Interior, Louisville, Kentucky, hereafter cited as Zehner, NM, 1976.

67. Meyer, PD, 1976, p. x.

68. Krey and Hardy, RFP; McClendon, SRP; and Pinder, "Field Study."

69. DeBuchananne, "Statement," p. 137.

70. KSTC, 1972, p. 8.

71. Meyer, PD, 1976, p. 44; and Zehner, NM, 1976, p. 270.

72. Rhodes, APS.

73. K. Knoll, "Reactions of Organic Wastes and Soils," BNWL-860 (Washington, D.C.: U.S. Atomic Energy Commission, 1969).

74. P. Neubold, Absorption of Plutonium-239 by Plants, ARRCL-10 (GB Arg. Res. Council, 1963); P. Neubold and E. Mercer, "Absorption of Plutonium-239 by Plants," ARRCL-8 (GB Agr. Res. Council, 1962); and J. Rediske et al., "The Absorption of Fission Products by Plants," HW-36734 (U.S. Atomic Energy Commission, 1955).

75. V. Hale and A. Wallace, "Effects of Chelates on Uptake of Some of Heavy Metal Radionuclides from Soil by Bush Beans" *Soil Science* 109 (1970) 26ff. See also K. R. Price, "A Review of Transuranic Elements in Soils, Plants, and Animals," *Journal of Environmental Quality* 2, No. 1 (1973): 62.

76. E. Romney et al., "Persistence of Plutonium in Soil, Plants, and Small Mammals," *Health Physics* 22 (1970) 487ff.

77. C. E. Christenson and T. R. Thomas, "Movement of Plutonium Through Los Alamos Tuff," in *Second Ground Disposal of Radioactive Wastes Conference*, TID-7628 (Chalk River, Canada: U.S. Atomic Commission, 1961). See also Meyer, PD, 1976, pp. 45–46.

78. Weiss and Columbo, EIM, pp. xxii–xxiii, 121–35.

79. Ibid., pp. 121, 134.

80. B. F. Hajek, "Letter to H. G. Holton of NECO, February 9, 1976," in U.S. Congress, 1976, p. 274.

81. Laudan, SV, p. 77.

82. U.S. Environmental Protection Agency, *Report to Congress on Hazardous Waste Disposal* (Washington, D.C.: Government Printing Office, 30 June 1973), pp. v, 10, 87.

83. Meyer, PD, 1976, p. 51.

84. H. H. Zehner, hydrologist, USGS, *Preliminary Hydrological Investigation of the Maxey Flats Radioactive Waste Burial Site, Fleming County, Kentucky*, USGS Open-File Report 79-1329, Louisville, Ky., 1979, p. 2.

85. H. H. Zehner, *Hydrological Investigation of the Maxey Flats Radioactive Waste Burial Site, Fleming County, Kentucky*, USGS Open-File Report 83-133, Louisville, Ky., 1983, pp. 3, 132.

86. Polluck and Zehner, CA, 1981, p. 7.

87. Ibid.

88. Ibid.

89. Cicero, *De Legibus,* 3., 3.

90. McMullin, VIS, sec. 2.

91. N. Rescher, "The Ethical Dimension of Scientific Research, in Klemke, Hollinger, Kline, IRPS, p. 247.

92. I would like to thank Steven Goldman for formulating this and the following objection.

93. Scheffler, SS, pp. 2–3.

94. See Scriven, JV, pp. 219–47; for a defense of this position on rationality and methodological rules, see K. Shrader-Frechette, "Scientific Method and the Objectivity of Epistemic Value Judgments," in *Proceedings of the Eighth International Congress of Logic, Methodology, and Philosophy of Science*, ed. J. Fenstand, I. Frolov, and R. Hilpinen, Studies in Logic and the Foundations of Mathematics (Amsterdam: North-Holland Publishing Company, forthcoming 1988), hereafter cited as Shrader-Frechette, EVJ.

95. I. Scheffler, "Discussion: Vision and Revolution: A Postscript to Kuhn," *Philosophy of Science* 39 (1972): 369.

96. See K. Popper, *The Open Society and Its Enemies,* (London: Routledge and Kegan Paul, 1945), 2: 205–208; and Scheffler, SS, pp. 68, 72–73, 86–88.

97. Scheffler, SS, p. 124.

98. See Shrader-Frechette, EVJ, for a defense of this position.

99. This is discussed at length in Richard Meehan's *The Atom and The Fault* (Cambridge, Mass.: MIT Press, 1984).

100. For a more extensive analysis of free, informed consent and its relevance to technological risk, see K. Shrader-Frechette, *Risk Analysis and Scientific Method* (Boston: Reidel, 1985), chaps. 4–5, hereafter cited as Shrader-Frechette, RASM.

101. For a more elaborate due-process argument, see K. Shrader-Frechette, *Nuclear Power and Public Policy,* 2d. ed. (Boston: Reidel, 1983), chap. 4 (for equity arguments related to this point, see chap. 2); K. Shrader-Frechette, *Science Policy, Ethics, and Economic Methodology* (Boston: Reidel, 1984), chaps. 5, 7; and Shrader-Frechette, RASM, chap. 3.

102. J. Rawls, *Theory of Justice* (Cambridge: The Belknap Press of Harvard University, 1971), pp. 152–155.

Part III
A Critique of the Idea of Progress

The Idea of Progress in Our Time

CHRISTOPHER LASCH

The history of the twentieth century does not appear, at first glance, to give much support to the idea of progress—to the idea, that is, that things are not only getting better but that the historical pattern of improvement is cumulative and irreversible. It was the irreversibility of improvement that defined the belief in progress so widely shared in the eighteenth and nineteenth centuries. Impressive additions to the store of human knowledge seemed to have given mankind a new mastery over the conditions of its existence, and this knowledge, however unwelcome in some quarters, however disturbing to the authorities or subversive of older habits of dominance and submission, could not be erased from the record. Nor was there any reason to believe that the production of new knowledge would come to an end. On the contrary, everything pointed to a steady enlargement of the sum of knowledge, with its almost unlimited potential for liberation. "I see no limit," said T. H. Huxley in 1893, "to the extent to which intelligence and will, guided by sound principles of investigation, and organized in common effort, may modify the conditions of existence, for a period longer than now covered by history." Huxley was by no means the most unrestrained exponent of progress in his time. He dismissed the hope of "attaining untroubled happiness" as an illusion. "The majority of us," he insisted, "profess neither pessimism nor optimism." Yet he foresaw enormous material improvements, and moral improvements as well—the assertion of "intelligence" over the "instincts of savagery in civilized men."[1]

World War I, an unexpected regression to savagery on a global scale, for a time appeared to give the idea of progress a blow from which it could hardly recover. In 1920, Dean William Ralph Inge devoted his Romanes lecture at Oxford—the very same forum to which Huxley had addressed his "Evolution and Ethics," twenty-seven years before—to an attack on the dogma of progress, a "superstition," as he called it, that was "nearly worn out."[2] In *The Decline of the West,* published in the same year though completed in 1914, Oswald Spengler announced that the end of Western civilization was "obligatory and insusceptible of modification," adding that anyone who failed to see the twilight of the West "must forego all desire to comprehend history, to live through history, or to make history."[3] To members of the so-called lost generation, the war showed that "man . . . is selfish, irrational and unwittingly absurd," as an American journalist put it.[4] "After the colossal follies of the twentieth century, what sensitive and thoughtful person can believe in the natural goodness of man?" Even before the war, the intensification of international rivalry and class conflict had raised doubts about progress. "We might as well make up our minds," wrote George Bernard Shaw in *Man and Superman,* "that Man will return to his idols and his cupidities, in spite of all 'movements' and revolutions, until his nature is changed. . . . We must therefore frankly give up the notion that man as he exists is capable of net progress."[5]

Those who still believed in progress did not fall silent, of course, but a defensive, irritable tone crept into their pronouncements, formerly so unshakably confident. In 1932, Charles A. Beard introduced a collection of essays, *A Century of Progress,* with an attack on the "critics and scoffers" who questioned the "steady improvement of the lot of mankind in this world." Beard devoted more attention to critics of progress than to progress itself. "Writing under soft lamps or lecturing for fees to well-fed audiences, in comfortable rooms electrically lighted," these critics advocated a "return to agriculture and handicrafts." What they really objected to, Beard decided, was the prospect of sharing their highbrow culture with the masses.[6] Equating criticism of progress with European criticism of American democracy, Beard sidestepped the question of whether modern technology, by vastly increasing the capacity for destruction, had not canceled out many of the gains it made possible. The evasiveness of his defense of progress invited ridicule. Lewis Mumford, reviewing Beard's anthology for the *New Republic,* found the book "without distinction and point. And no wonder: for Progress is the deadest of dead ideas"—the "one notion that has been thoroughly blasted by the facts of twentieth-century experience."[7]

In the depths of the Depression, with democracy collapsing in Europe and the American economy in ruins, most readers of the *New Republic* would probably have agreed with Mumford that Beard's celebration of progress was hopelessly out of date, as anachronistic and empty as the rhetoric in praise of the belief in unlimited progress with which he concluded *The Rise of American Civilization:* "It meant an invulnerable faith in democracy, . . . in the efficacy of that new and mysterious instrument of the modern mind, 'the invention of invention,' moving from one technological triumph to another, . . . effecting an ever wider distribution of the blessings of civilization, . . . conjuring from the vasty deeps of the nameless and unknown creative imagination of the noblest order, subduing physical things to the empire of the spirit . . ."[8] Believers in progress, confronted with another global crisis they had not foreseen and could not explain, bereft of arguments, had nothing left but hot air. Yet hot air, if that is what it was, soon melted skepticism and brought about a remarkable change in the climate of opinion, a pseudosummer or Indian summer that lingers even today. After

World War II, the idea of progress came back into favor. This surprising revival of a discredited article of faith came about, one suspects, not because the human prospect had improved in any important way but because, in the deepening crisis of the twentieth century, men and women found themselves unable to get along without it. "The world today believes in progress," Sidney Pollard flatly declared in 1968. After devoting a whole book to a half-hearted and unconvincing attempt to show that such a belief is justified, Pollard revealed more than he intended when he concluded, "Today, the only possible alternative to the belief in progress would be total despair."[9] A doctrine solidly rooted in the material and political advances of the eighteenth and nineteenth centuries and shaken by the calamities of the early twentieth century thus gained new life, or at least a semblance of life, from the very disasters—the Final Solution, Stalin's Gulag, Hiroshima, the nuclear arms race—that ought to have finished it off once and for all. Pessimism, it seems, is a luxury we can no longer afford. The need to hope against hope, to believe (like Tertullian) even when belief has become absurd, explains why the true believers continue to berate their critics with more energy than they bring to the defense of progress itself. If progress is more than a "moral intuition," as Warren Wagar calls it, it ought to be possible to cite a historical record of steady improvement, one that can convincingly be projected into the distant future. Instead, we hear impassioned warnings against despair, fervent denunciations of those who would destroy our faith in humanity. The conclusion of Wagar's book, published in 1972, falls back on the same kind of rhetoric so congenial to Beard. "Let us not resort to the contemptible *Schadenfreude* of the neo-Augustinian theologians, the obscurantists, and all the pious and aesthetic and mystical refugees from progress who detect a fatal moral insufficiency in the heart of man."[10] But what if historical experience forces us to postulate exactly such an insufficiency? Don't we have an obligation to come to terms with that experience? Perhaps it might help us not so much to resolve the interminable debate about progress as to explain why it will always remain inconclusive.

The postwar revival of progressive ideologies, after all, did not settle the issue; it merely made clear what should have been clear all along, that the question of progress in its existing form cannot be settled by an appeal to historical evidence, whatever that evidence may tell us about the "moral insufficiency in the heart of man." Wagar's qualified optimism, by no means unreasonable in itself, sums up the case for progress in our time; but the case is hardly conclusive. "Despite interludes of totalitarian tyranny, the mass of men enjoy a better life, perhaps, than was ever before possible."[11] The trouble with such statements is not necessarily that it is impossible to agree on definitions of a better life, but that the evidence on this point is so ambiguous. Even if we could overlook the poverty and famine to which millions are doomed, not to mention the threat of overpopulation, exhaustion of nonrenewable resources, or the threat of nuclear or ecological disasters, we would still find it difficult either to defend or to refute sweeping comparisons between existing conditions and the conditions supposedly endured by the "mass of men" in the past.

The debate about progress turns on the juxtaposition of vague, stylized images of modern society with starkly contrasting images of something called "traditional" society. Skeptics argue that if a better life for the masses includes collaboration with nature, continuity with ancestors and descendants, and a sense that a human life has a place in some larger scheme of things, the modern age marks no improvement over the past. Defenders of modernity reply that the anonymity of urban life provides privacy against intrusive neighbors and a range of personal choice unthinkable in the preindustrial village. What skeptics see as the stability of premodern society appears to the party of hope as stagnation. Our much-heralded emancipation from tradition, on the other hand, so obvious to the hopeful, looks to the skeptical like *anomie* and alienation. This ideological battle of clichés has settled into a predictable sequence of moves and countermoves; no doubt the familiarity of these strategies, together with the inconclusive character of the fighting, constitutes a large part of its appeal as a staple topic of discussion. It is a battle in which neither side suffers much damage. The argument about progress can no longer be understood as an argument at all. It has become a cultural ritual that gives expression to our ambivalence about modern life but leaves undisturbed our underlying belief in improvement, or at least in the inevitability of technological development.

Instead of joining this ritual debate, I propose that we take a closer look at the assumptions behind it. It is because both sides share these assumptions that the debate remains inconclusive. Even in the twenties, when belief in progress seemed to have fallen to a low ebb, doubters found it difficult to escape from the subtle coercion of a progressive view of history. Take the case of Freud, whose insistence on the power of the irrational was so often invoked by those who now denied the reality of progress. It is a measure of the tenacity of progressive thought that it continued to influence even such a bold and unconventional thinker as Freud. His psychological discoveries pointed to the need to reconsider the intellectual foundations of liberalism; yet his own writings on society and culture rest on those same shaky foundations. In the two books that contributed most directly to the debate about progress, *Civilization and Its Discontents* and *The Future of an Illusion,* the latter an attack on religion, Freud argued that humanity's hope lies in extending the hold of reason over the passions. He clung to a second convention of liberal thought as well: that the gradual assertion of reason over appetite, the essence of civilization, can be likened to the individual's growth from infancy to maturity. Note the implications of this biological analogy. Because individuals eventually mature, no matter how reluctantly or incompletely, we can make the same inference about cultures. Although both individuals and society pay a price for maturity—a guilty conscience, in Freud's view—it enables people to live in relative peace with themselves and with their neighbors. To pine for the lost innocence of childhood, therefore, is foolish; it cannot be recovered and is based on illusions anyway.

At the cultural level, according to Freud, these illusions take the form of a belief in supernatural powers; and if our modern disillusionment deprives us of the childlike security of dependence, it is better that we learn to depend only on ourselves. Self-reliance, for individuals and for humanity as a whole, is the real mark of maturity. "Men cannot remain children for ever. . . . It is something, at any rate, to know that one is thrown upon one's own resources."[12] Freud's tone is wistful but firm: let us put away childish things. (The anthropologist Robert Redfield, who shared Frud's appreciation of the attraction of the primitive, the safety of the "little community," resorted to the same rhetoric: "I find it impossible to regret that the human race has tended to grow up.")[13] Because the resources on which mankind must now depend include science, there is no reason for despair. Science "has taught [men] much since the days of the Deluge and it will increase their power still further." Religion, Freud adds, is "comparable to a childhood neurosis," and man can expect to "surmount this neurotic phase."[14]

Elsewhere in *The Future of an Illusion,* Freud states the problem of progress in its classic form. "While mankind has made continual advances in its control over nature and may expect to make still greater ones, it is not possible to establish with certainty that a similar advance has been made in the management of human affairs."[15] The second part of this sentence contains the usual argument for modern "pessimism." But it is Freud's statement of the problem, not his apprehension that our self-control has not kept pace with our control over nature, that betrays the residual influence of nineteenth-century thought. This formulation contains a number of questionable assumptions, best exposed by an examination of the competing conclusions to which it usually gives rise. The more hopeful of these conclusions, in the conventional discourse on progress, is the theory of cultural lag: although our ethics, politics, and social organization lag behind advances in technology, they can be expected to catch up, because the same scientific habits of mind that inform modern technology will eventually inform our understanding of politics and mortality. The second conclusion is conventionally pessimistic: our knowledge of the social world may always lag behind our knowledge of nature, because it is hard to achieve the objectivity of the natural sciences when human appetites and interest begin to intrude themselves into any discussion. As Morris Ginsburg put it in 1953, scientific advance "has depended on the ways in which purely deductive thinking has been balanced by observation and experiment," and the social sciences suffer from the disadvantage that "their methodology is still uncertain and objectivity . . . more difficult to attain."[16]

We can reconstruct the assumptions behind these platitudes as follows.

First, emotions are at odds with intellect and interfere with clear thinking. Thought too often tends to be guided by our wishes and desires rather than by vigorous logic. Science overcomes wishful thinking, but social science finds this more difficult, although it should try to model itself on the physical sciences and to distinguish sharply between facts and values, truth and hopes. The most important premise in this series is the equation of emotion with desire, the implications of which will be explored in a moment.

Second, physical appetites are the root of desire. It is because human beings are driven by bodily needs and subject to bodily limitations that their understanding is so limited. Only by escaping from the body or by overcoming its effects on consciousness can humans arrive at understanding. Once intelligence frees itself from the prison of the body, however, it can grasp timeless truths and one day, perhaps, conquer scarcity, sickness, and even death itself, the ultimate indignity. Condorcet, the first modern thinker to state the idea of progress in its full-blown form, dared to hope that the "day will come when death will be due only to extraordinary accidents or to the decay of the vital forces," so that "ultimately the average span between birth and decay will have no assignable value."[17] George Bernard Shaw, who ridiculed the idea of progress, nevertheless envisioned the indefinite prolongation of life in *Back to Methusaleh,* a work notable for its celebration of disembodied intelligence—the intellectual basis of all prolongevity doctrines, indeed of the modern cult of technology in general. Shaw's ancients are "still not satisfied" even with the abolition of death. Moved by the "greatest of gifts, curiosity," they "press on to the goal of redemption from the flesh, to the vortex freed from matter, to the whirlpool in pure intelligence." They resent the "machinery of flesh and blood," which "imprisons us on this petty planet and forbids us to range through the stars."[18]

Third, knowledge is power—not an understanding of the limits of human power. Knowledge brings power and control not only over the body but over the animal kingdom as a whole and over the animal element in humans; and one important measure of progress, accordingly, is the distance human beings have traveled from animal origins. Carl Becker maintained that this was the only measure of progress uncontaminated by value judgments. It was pointless, in Becker's judgment, to ask "whether the human race is moving toward either a good or a bad end." But it was possible to ask "in what essential respects man has become different from what he was, from his cousins the apes." Posed in this way, the question of progress answered itself: whereas man once "associated on scarcely more than equal terms" with the apes, now he displayed them in zoos. "Man has increasingly implemented himself with power"; and "all that has happened to man in 506,000 years may be symbolized by this fact—at the end of the Time-Scale he can, with ease and expedition, put his ancestors in cages."[19]

It would be hard to find a better illustration of the species-centered point of view, as Bernard James calls it,[20] that informs the whole debate about progress and unites the various assumptions on which it rests. The pretense that our domination of nature provides an ethically neutral criterion of progress can scarcely stand up under analysis, because it is precisely the extension of human capacities, the perfection of human control over the conditions of human existence, that the modern world equates with freedom from want, fear, ignorance, and dependence—the highest values modern man is able to imagine. The commonplace reminder that freedom, like everything else, exacts its price; that freedom is a burden many would willingly throw off; that it deprives us of the security of prescribed roles; that it confronts us with a baffling array of choices; that it leaves us alone in the world without a heavenly father to fall back on; that the insecurity of the modern world gives rise to totalitarian movements that promise an "escape from freedom": none of this calls the value of freedom into question or challenges the premise that freedom, with all its attendant anxieties, is the ultimate measure of progress. A critique of the idea of progress cannot evade the ethical question of whether freedom, understood as human autonomy—human control over nature and over human nature as well—deserves the high regard in which it is held by the modern world.

To say that this is an ethical question does not mean that it is merely a "value judgment" on which there will be as many opinions as there are participants in the discussion. The question of whether we should set a high value on human autonomy cannot be disentangled from the factual question of whether human beings are capable of the kind of autonomy they aspire to or whether autonomy is not, in fact, largely an illusion, one that is even more pervasive and appealing than the illusion that a benevolent providence looks out for our best interests. If the belief in autonomy and freedom has no basis in fact, we will also need to know more about the emotional basis of this particular illusion.

It is an exaggeration, of course, to say that it has no basis in fact. Science and technology have unquestionably enhanced human control over nature; the point is that this control is still subject to severe limitations. The more control humans exert, the more striking those limitations become. Human interference with natural processes produces far-ranging effects that will always remain unpredictable in part, even in large part. This is the lesson of ecology, which confirms the ancient insight that even the best intentions often lead to unexpected and highly undesirable consequences. There is no need to rehearse the abundant evidence of man's bungling attempts to shape the natural environment according to the pattern of his desires. The chief finding of ecological science is well known and undisputed: that life on this planet constitutes an integrated system, modification of any part of which brings about incalculable effects on every other part. The staggering complexity of these interactions reawakens the emotion of awe and reminds us, incidentally, that intelligence and emotion, instead of working at cross purposes, depend on one another. It is only the equation of emotion with desire that leads to the disparagement of emotion on the grounds that it generates wishful thinking. But there is nothing wishful about awe. Indeed, it runs directly counter to our wish to believe ourselves masters of all we survey. Yet we cannot escape the impression that fear and trembling are a more appropriate response to the facts at our disposal than a heightened confidence in our own powers. Our triumph over the animal kingdom turns out to be far more modest than we had imagined. "Is human life to be interpreted more in terms of continuities with nature or in terms of distinctiveness from nature?" Mounting evidence tilts to the former position and thus leads to a new respect, as James Gustafson points out in his *Ethics from a Theocentric Point of View,* for the limits of freedom. This evidence consists of an understanding not only of ecological systems but of the vastness of the universe, the relative insignificance of our place in it, and the eventual extinction of life on earth not through human agency but through the inevitable working of time. These facts, Gustafson writes, undermine the "traditional Western assurance that everything has taken place for the sake of man."[21]

Having briefly considered the scientific evidence that undermines a belief in human autonomy, we need to consider the emotional source of that belief. If it is not justified by the evidence, why do humans nevertheless continue to believe in their capacity for mastery and freedom? The answer is that this belief serves an important emotional need, the need to deny our dependence on forces we cannot control. It is ironic that Freud, whose clinical investigations did so much to weaken the illusion of psychic freedom, in his speculations on culture nevertheless adhered to the conventional view that religion recreates the emotional security of infancy, the "oceanic" contentment that recedes with maturity. Freud's discoveries about unconscious mental life, supplemented by those of later investigators, indicate that the problem is more complicated than that. It is precisely the infant's discovery of his dependence on caretakers who cannot be depended on to gratify his every desire that threatens the oceanic feeling of oneness and peace. That feeling of oneness, moreover, springs not from a comforting awareness of dependence on "higher powers" but, on the contrary, from the infant's early illusion of omnipotence, his inability to distinguish his own wishes from the world around him. When he does learn that the world does not exist for the sole purpose of gratifying his desires, he tries to recapture the illusion of oneness either by imagining that his own being melts into his surroundings or by making himself absolutely self-sufficient.

It is true that religion sometimes encourages the first of these strategies of evasion and denial, but it is also true that religion can subject both strategies to relentless criticism. The deepest religious message is the injunction to love life even though it is not organized around the fulfillment of human wishes. Religion urges us to accept our dependence on uncontrollable forces not as a source of despair but as the condition of our being—as such, the source of whatever happiness we can expect to enjoy. According to Gustafson, the religious attitude is best described as "dependence on, and respect and gratitude for, what is given."[22] From this point of view, the replacement of religion by science does not look much like maturity, as Freud and so many others have seen it. Insofar as science associates itself with the denial of dependence, ultimately with the denial of death, and with the imposition of human will on nature, it looks like a regression to the second of the strategies of denial just alluded to—an attempt to preserve the illusion of our self-sufficiency and to ward off the unwelcome awareness of our dependence on powers unamenable to human will. The conventional pessimism that attributes our modern uneasiness to lost innocence proves no more tenable than the optimistic celebration of modern technology. In a culture that still clings to the infantile illusion of self-sufficiency, maturity is a myth. Although it is vaguely consoling to think of ourselves as sadder but wiser than our ancestors, only the first of these adjectives, alas, describes the modern condition.

I hope I have said enough to convince you not that the pessimists about progress are right, but that the whole debate between pessimism and optimism is misconceived. It is pointless to argue about whether the present age is better than the past and equally pointless to try to resolve the dispute by making the obvious point that in some ways it is better and in some ways worse. This kind of historical bookkeeping, which balances gains against losses, amounts to an exercise in futility. The controversy about progress draws so heavily on the biological metaphors of youth, maturity, senescence, and rejuvenation that it inevitably assumes just what needs to be demonstrated: that the accumulation of experience necessarily makes for wisdom. It makes for wisdom only if we bring to the interpretation of experience a spirit of humility and contrition.

Unfortunately the arrogance of the modern age remains undiminished by our recent history of disasters, a history with plenty of precedents, to be sure, in past disasters. Even a sober reassessment of our recent history leads all too quickly and predictably to the question of whether humanity stands on the brink of suicide. As Bernard James observes in *The Death of Progress,* the question assumes that "man, this strutting, puffing, assertive biped, is the measure of all things, the final marvel of creation, his extinction the ultimate philosophic issue."[23] The real question is not whether recent disasters portend collective suicide but whether the natural ills that flesh is heir to, the inevitable afflictions and decay the knowledge of which gnaws at us even in the best of times, precludes gratitude and hope, a joyful affirmation of the rightness of a world that was not created for our particular benefit. The question is whether hope must always be tied to the prospect of improvement, the prospect that somewhere, somehow, humanity can find deliverance, in George Bernard Shaw's parody of St. Paul, "from the body of this death." Hope is not simply or even primarily an attitude toward the future. It is better understood as trust in life itself, an underlying disposition to see the promised land not as a distant objective but as a present reality, the ground and basis of our being. This kind of hope refuses to be "frightened by the prospect of doom on all man's works," as Richard Niebuhr once observed.[24]

In the face of doom, it remains "not despairing but confident." Hope that depends on the modern myth of progress, on the other hand, can sustain itself only by the willful exercise of wishful thinking—not a very solid foundation on which to build a better life.

NOTES

1. Quoted in W. Warren Wagar, *Good Tidings: The Belief in Progress from Darwin to Marcuse* (Bloomington: Indiana University Press, 1972), p. 49. Far more sweeping assertions of the nineteenth-century faith in progress can easily be cited. In his *Social Statics* (1850), Herbert Spencer declared: "The ultimate development of the ideal man is logically certain. . . . Progress, therefore, is not an accident, but a necessity. . . . It is certain that man must become perfect. (Quoted in Sidney B. Fay, "The Idea of Progress," *American Historical Review* 52 [1947]: 237). John Fiske echoed these sentiments in *The Destiny of Man Viewed in the Light of His Origin* (1884): "So far from degrading Humanity, the Darwinian theory shows us distinctly for the first time how the creation and the perfecting of Man is the goal toward which Nature's work has all the while been tending." (Quoted in Clarke A. Chambers, "The Belief in Progress in Twentieth-Century America," *Journal of the History of Ideas* 19 [1958]: 200–201.)

2. Quoted in Morris Ginsberg, *The Idea of Progress: A Revaluation* (London: Methuen, 1953), p. 11.

3. Quoted in Sidney Pollard, *The Idea of Progress* (New York: Basic Books, 1968), pp. 164–65.

4. Frank Snowden Hopkins, "After Religion, What?" (1934), quoted in Chambers, "Belief in Progress," p. 204.

5. George Bernard Shaw, *Man and Superman* (London: Constable, 1929), p. 206.

6. Charles A. Beard, ed., *A Century of Progress* (New York: Harper, 1932), pp. 9 and 16.

7. *New Republic* 76 (September 6, 1933): 106.

8. Charles A. Beard and Mary R. Beard, *The Rise of American Civilization*, vol. 2 (New York: Macmillan, 1930), p. 831.

9. Pollard, *Idea of Progress*, p. 203.

10. Wagar, *Good Tidings*, pp. 355–56.

11. Wagar, *Good Tidings*, p. 240.

12. Sigmund Freud, *The Future of an Illusion* (New York: Norton, 1961), pp. 49–50.

13. Quoted in Wagar, *Good Tidings*, p. 316.

14. Freud, *Future of an Illusion*, pp. 43, 50, and 53.

15. *Future of an Illusion*, p. 7.

16. Ginsberg, *Idea of Progress*, pp. 52–53.

17. Quoted in Thomas A. Spragens, Jr., *The Irony of Liberal Reason* (Chicago: University of Chicago Press, 1981), p. 56.

18. George Bernard Shaw, *Back to Methuselah* (London: Constable, 1922), p. 260.

19. Quoted in Charles Van Doren, *The Idea of Progress* (New York: Praeger, 1967), pp. 226–27.

20. Bernard James, *The Death of Progress* (New York: Knopf, 1973), pp. 32 and 57.

21. James M. Gustafson, *Ethics from a Theocentric Point of View,* vol. 1 (Chicago: University of Chicago Press, 1981), p. 83.

22. Gustafson, *Ethics,* vol. 1, p. 61.

23. James, *Death of Progress,* p. 57.

24. H. Richard Niebuhr, *The Kingdom of God in America* (Chicago: Willett, Clark, 1937), pp. 26–27.

Science, Technology, and the Theory of Progress

CARL MITCHAM

"Scientific progress" and "technological progress" are theories that, at first sight, appear to refer to unproblematic phenomena. Isn't twentieth-century science clearly superior to medieval or Greek science? Would anyone want to give up the modern making and manipulating of artifacts for premodern technics? Aren't the technologies of medicine, transportation, and communication, obviously better than they have ever been?

Despite this apparent obviousness, there exists a wealth of literature on the idea or theory of progress. This literature includes (first) eighteenth-century Enlightenment arguments for the theory in the face of what was construed as premodern prejudice against it, (second) nineteenth-century Romantic criticisms of the theory as based on false and limited notions of human nature, (third) twentieth-century analysis of the historicophilosophical development of the theory, studies of the historicosocietal presuppositions and implications of the theory, and continuing exchanges between representatives of the Enlightenment and Romantic traditions of *apologia* and *judicium*. Bibliographer though I am, I will resist the temptation to cite this literature in any detail, but its wealth should occasion pause. Surely its existence alone could give pause and rise to some wonder about whether the theory is really as unproblematic as it initially appears.

The body of literature in question has, however, another characteristic closely related to its volume—a certain lack of conclusiveness. Within this wealth of literature, and no doubt partly the cause of it, certain themes and issues regularly recur. What are the relevant and irrelevant aspects of historical change? What is the proper role of science and technology in human affairs? What is the relation between science-technology and human nature? Are scientific and technological change determinative of other kinds of change? Are scientific and technological change related? Is either always cumulative? The discussion of the theory of progress itself exhibits only limited progressiveness. Indeed, I take it that the theory of progress is, despite the literature, fraught with philosophical issues still to be explored.

Many philosophers, for instance, more or less assume the commonsense notion of the obviousness of progress in science and technology. Larry Laudan's widely discussed *Progress and Its Problems* (1977) is a good case in point. In criticizing the philosophies of science associated with Thomas Kuhn, Imre Lakatos, and Paul Feyerabend, Laudan proposes to invert "the presumed dependency of progress on rationality" (in Kuhn, Lakatos, and Feyerabend) and "to show that we have a clearer model for scientific progress than we do for scientific rationality; that, moreover, we can define rational acceptance in terms of scientific progress." In short, Laudan argues "that *rationality consists in making the most progressive theory choices,* not that progress consists in accepting successively the most rational theories" (p. 6, his emphasis). Philosophies of science (such as Kuhn's or Feyerabend's) that wind up calling progress into question on the basis of their models of rationality are thus prima facie to be doubted because they use the less well founded to criticize the more well founded. The burden of Laudan's book— despite what might be thought implied by its title—is to deny that progress is itself a problem. Instead, by analyzing the various kinds of empirical and conceptual problems solved by science, Laudan proposes to show how it obviously progresses and thus increases in rationality.

Without directly addressing Laudan's claims, my argument is that despite its apparent obviousness, progress is truly a philosophical problem; here as in many other cases, appearances can be misleading. There are basic questions to be raised about the theory of progress, especially in relation to science and technology. My argument, however, will offer no solutions. In an effort to eschew that rhetoric to which the literature on progress so readily lends itself, I will stress conceptual distinctions and simply struggle to begin identifying the key questions—and some of the ways these have been unreflectively answered.

1.

One of the most crucial things to notice is that our initial judgment of the relation between past and present science and technology as one of obvious progress could under certain easily imaginable circumstances just as obviously be reversed. Indeed, the rise of this imaginative reversal, if you will, is a characteristic phenomenon of twentieth-century intellectual history. If tomorrow our military technology should, despite all efforts to the contrary, suddenly produce a worldwide conflagration, popular opinion—what might be left of it—would certainly be that somehow our "technological progress" had been a vast mistake, and was in reality no progress at all. Increased and increasing risks and unwonted side effects of technological change in the forms of environmental pollution (from acid rain to ground-water contamination), iatrogenic disease (new forms of staphylococcus infections resistant to antibiotics), ecological disaster (ozone depletion), and the risks of artifice and its behavioral correlates (from nuclear weapons to toxic chemicals and AIDS) all harbor the possibility of such reversals in judgment. Whether what we currently experience as progress really is change for the better is dependent to some extent on where this change leads in the future.

The issue here is related to that of Aristotle's discussion of when *eudaimonia* is properly attributed to a person *qua* person—whether this can be done at any point short of death or perhaps not even then (*Nicomachean Ethics* 1. 10). It also broaches hermeneutic issues at the fundamental level of self-understanding or self-interpretation. The primordial manifestation of the good is as that which is our own. The touch and feel of our own bodies, the taste and smell and feel of those foods and clothes and homes with which we were raised, even the emotional relationships in our families, are all prereflectively taken to be good and preferred to other tastes, smells, and emotions—even when it can be objectively shown (and seen) that others are better. This identification of the good with what is our own readily expands into a prizing of our current state of affairs and a conservative resistance to change. It further prolongs itself into a prereflective love for the ancestral, that from which our own has come. We have to work at, struggle for, an emancipation from the prereflective identification of the good with what is our own and our own tradition that will allow us to choose a different good or the good in a less qualified sense.

In a highly scientific and technological society such as ours, then, the belief that modern science and technology are good—good even in their tendency to bring about future changes—is itself a version of this initial prereflective identification of the good with what is one's own. Only on the surface is there an opposition between traditional conservatism and the typically modern commitment to change. The commitment to change, no matter what the costs, is the distinctive way of being in the world that is our own, and thus naturally tends to be affirmed and defended. The commitment to change has become in fact the conservative position. We naturally resist that most basic of changes, which would be to delimit or to restrict change. The theory of progress justifies such resistance, paradoxically, rationalizing the status quo.

The historiography that creates, from Democritus to Galileo, a tradition of anticipations of modern science is but the extension of such prereflective identification of the good as change in the form of scientific and technological progress. At the same time, so much does the prereflective identification of the good with one's own influence the public reading of history, that were there to occur a catastrophe attributable to scientific or technological development, this itself could be expected eventually to take on the character of the good—good, perhaps, in bringing about a destruction of the illusions of science, a release from the evil forces of technology, or a freedom from misleading technological dependence. This alternative identification of the good would lead in turn to an alternative interpretation of the past and a prizing of aspects that are consonant with this new view—among others, premodern skepticism about progress and the Romantic criticism of science and technology.

2.

Against the background of such a possible reversal of understanding, it becomes incumbent upon us to inquire further into the essential character of progress. Let me begin with some rather simple and obvious distinctions about different kinds of historical change, noting that only one of these constitutes progress in any unqualified sense, and then suggesting that this sense is not likely to be found in history in any unqualified way.

Not all motion or change is historical change. Water flowing down hill, human breathing, the rising and setting of the sun—none of these natural changes would normally be considered historical in character, nor (except in the weakest of senses) could they be called progressive. Yet it is not all that easy to specify the difference between historical and natural change. Whereas once (and still to some extent) planetary motion was considered nonhistorical, now many argue that the solar system does have a history, of which annual trips around the sun constitute an element. This implies, of course, that the dividing line between historical and nonhistorical change might itself be historical, or that what we think of as historical change is at bottom natural, or that historical change is the most all-inclusive kind of change. In the present instance, however, it can only be noted that there exists an unclarified relation between these two types of change. It is also not clear whether mechanical or artificial change—such as that exhibited by a clock or steam engine in operation or an automobile being driven from one point to another—is best interpreted in relation to nature or to history.

Be that as it may, with regard to historical change, there are five different theories about its basic character. Historical change can be construed as circular or cyclical; primarily circular or cyclical, but with some one important difference introduced at a specific point in time; progressive; regressive; neither circular nor progressive nor regressive, but ambiguous.

For the purposes of our discussion, let us symbolize these different theories about historical change:

T indicates some state of affairs at a given time.

Subscripts 1, 2, 3, and so forth indicate the temporal priority or relationship between any two such states of affairs.

Superscripts a, b, c, and so forth indicate the human activities or achievements or nonhuman physical characteristics of these states of affairs.

With this symbolization, the following schemata can represent each of the five theories of historical change.

(a) The circular theory:
$$T_1{}^a \longrightarrow T_2{}^b \longrightarrow T_3{}^a \longrightarrow T_4{}^b \ldots$$

This is probably the most common theory of the character of human experiences as exemplified by the cycles of day and night, springtime and harvest.

(b) The circular theory with a difference:
$$T_1{}^a \longrightarrow T_2{}^b \longrightarrow T_3{}^{a+c} \longrightarrow T_4{}^{b+c} \longrightarrow T_5{}^{a+c} \ldots$$

Note that the addition of the superscript c can be in the form of either a regression or a progression. The Biblical account of the Fall is a prototype of the former, the theology of the Incarnation illustrates the latter.

(c) The progressive theory:
$$T_1{}^a \longrightarrow T_2{}^{a+b} \longrightarrow T_3{}^{a+b+c} \ldots$$

This theory is most generally thought to be manifested by changes in science or in technology. Note, however, that in one sense it merely expands into a universal process one moment in theory (b)—that of T_3—while in another sense it could easily be combined with theory (a) to yield a much richer interpretation of historical reality. In fact, it is some version of this second possibility that is most usually proposed by the progressive theory.

Two refinements of the progressive theory are what might be called the perfective and the defect-reductive theories. In the perfective theory of progress the cumulation of a + b + c . . . is presumed to be reaching toward some point of perfection. This was, for instance, the early modern view of scientific progress. By contrast, in the defect-reductive theory of progress the a + b + c . . . are thought of as negating certain defects.

(d) The regressive theory:
$$T_1{}^{a+b+c} \longrightarrow T_2{}^{a+b} \longrightarrow T_3{}^a \ldots$$

This theory just turns theory (b) around. It is readily illustrated by something like Plato's theory of the decay of the virtuous state in the *Republic*, Book 8.

(e) The theory of ambiguity:
$$T_1{}^a \longrightarrow T_2{}^b \longrightarrow T_3{}^c \ldots$$

This theory can, of course, readily be combined with any one of the other four. Most commonly, perhaps, it is combined with theory (a) to yield Shakespeare's "sound and fury, signifying nothing" (*Macbeth*, act 5, scene 5).

In light of this conceptual framework, one can make the following general observations about change in science and in technology.

The Cyclical Theory

Although no one explicitly argues a pure cyclical theory of scientific or technological change, some form of recurring activity is clearly at the base of many discussions: the scientific method is applied over and over; scientific revolutions repeat certain cycles of initiation and consolidation (Kuhn); technological innovation exhibits regular patterns of development.

The Cyclical Theory with a Difference

Virtually all historical accounts of the origin of modern science (and to some extent modern technology) attribute it to western Europe during the sixteenth and seventeenth centuries. Before this there were cycles of birth and decay in science, afterward there have been more cycles of birth and decay—but with a difference. Kuhn explicitly adopts such a view in his distinction between preparadigmatic and paradigmatic science.

The Progressive Theory

The progressive theory is, as already indicated, the most common interpretation of change in science and technology, but it is important to note that there is some difference of opinion about exactly what kind of progress is taking place in each instance. For some (who often point to discontinuities in the cumulation of techniques and skills), it is information or knowledge—that is, scientific knowledge; for others (who are likely to draw attention to conceptual discontinuities between theories and paradigms), it is various devices and powers. There are also debates about whether progress is of the perfective kind (the original eighteenth-century view, which thus presumed progress would eventually reach some kind of terminus) or of the defect-reductive kind (a view associated with contemporary philosophers of science such as Karl Popper and Kuhn).

The Regressive Theory

The regressive theory tends to be characteristic of humanist criticisms of science and technology. One exemplary exponent of this theory is Jean

Jacques Rousseau who, in his "Discourse on the Arts and Sciences," argued that advancements in technology brought about a decline in virtue. Lewis Mumford likewise maintains that since the invention five thousand years ago of rigid hierarchical forms of social organization that he calls the "Megamachines" there has been a progressive mechanization of human life, which is in reality a kind of repression from the rich variety of primitive polytechnics.

Lest one think such a view is only to be associated with humanist critics of science and technology, however, it can be noted that Nicholas Rescher has recently put forth a highly qualified regressive theory. Against the background of what he presents as accelerations in scientific progress over the last two hundred years, Rescher argues that there is a slowing down or deceleration—primarily because scientific breakthroughs are becoming more and more expensive. The advanced industrial countries will simply not be able to sustain the capital investments in research necessary to prolong the recent rates of scientific discovery.

The Theory of Ambiguity

The theory of ambiguity has been proposed in a number of different forms by Jacques Ellul, Stephen Toulmin, and others. Whenever anyone admits that scientific or technological change has brought with it certain costs (or risks) as well as benefits, there is an implicit admission of some ambiguousness— that is, a lack of any simple cumulativeness—in historical development. Usually, however, the theorists in question want to subordinate this ambiguity either to a larger-scale regression or to a larger-scale progression.

3.

Knowledgeable individuals can make reasonable cases for quite different theories. Which is correct? How is it that good examples from the history of science and technology can be found to support or justify each theory? Or, to reverse the question: why does each theory turn out to be useful under different circumstances for understanding the history of science or technology?

Judgments about the applicability of any one of these theories to particular historical circumstances are dependent on at least three factors. The first is the time frame or the length of time to be considered. This might conveniently be termed the backward-and-forward or longitudinal slice of reality. To apply theory (a) to the day-night cycle, for instance, one would have to take into account at least a thirty-six-hour period of time.

The second factor could be called the lateral slice, referring to the location of the relevant action or activity. For the day-night cycle, the lateral slice of the solar system would need to be limited to the earth. In considering the time slice from, say, 1900 to 1950, one could focus on the lateral slices of France, England, or India and arrive at very different judgments about the character of historical change in each case.

A third factor could be called a vertical slice and would indicate, for example, the aspect of French, English, or Indian culture being dealt with in the period under consideration. Historical changes in the area of art might well fit one theory, those in the area of politics quite another.

In light of this multidimensional matrix, one can readily suggest that progress in any unqualified sense is not likely to be found in history—perhaps not even in scientific or technological history broadly conceived. History is almost always going to be more complex than that.

4.

Coordinate with the identification of these three factors, certain observations can be made about the formal relations between alternative theories. First, theory (b), the circular theory with a difference, involves combining (a) and either (c) or (d). Second, theory (c), the theory of progress, and (d), the theory of regress, are but mirror images of each other. As a result, only theories (a), circular, (c), progressive, and (e), ambiguity, stand out as calling for more extensive formal analysis.

Taking into account another formal factor plus a material one, the direction of further analysis can be still more sharply focused. Theory (a), the circular theory, is not only formally compatible with both theory (c) and theory (e), it is in fact commonly accepted and assumed—although not always with full explicitness—as a material precondition for the phenomena of progress or ambiguity. Hence, it appears that there are primarily two alternatives requiring further analytic study, (a) and (c)/(e).

To facilitate this further analysis, the schemata of progress and ambiguity given in (c) and (e) can be rearticulated as follows. Progress can be crudely defined as change or difference over time plus betterness. Ambiguity requires only change or difference over time. Let us slightly modify the symbolism:

T_1 and T_2 are two states of affairs.

The sequence from T_1 to T_2 can be termed progress if and only if it meets the following four minimal conditions:

(a) T_2 and T_1 are in some way related.
(b) T_2 is different than T_1.
(c) T_2 comes after T_1 in time.
(d) T_2 is better than T_1.

When such a relationship obtains it can be summarily indicated by
$T_1' \longrightarrow T_2''$.
The line —— indicates relationship.
The arrow stands for temporal sequence.
The prime marks ' and '' symbolizes betterness.

If for some reason condition (d) fails—and one should recognize the difficulty of clearly specifying "betterness"—then the relationship

$T_1 \longrightarrow T_2$

will simply be ambiguous. Historically, the questioning of condition (d) either on empirical or conceptual grounds has led precisely to this result. The key question thus appears to be that of betterness and how it can be determined.

Stephen Toulmin's argument for replacing the theory of progress with a theory of evolution in science readily illustrates such a shift. As Toulmin presents it in *Human Understanding* (1972), the theory of evolution can involve two separate ideas: the *fact of descent* and the *doctrine of progression*. Darwin (and modern biology) is committed to the former but not to the latter.

> In contrast to all previous evolutionary theory, Darwin's populational theory of speciation accepted the hypothesis of descent, but was neutral about any supposed overall direction of historical development. . . . So far as the Darwinian mode of explanation went, all evolutionary changes must be accepted for what they were, in relation to their own ecological contexts. (p. 333)

Following Darwin's alleged lead, Toulmin, "instead of speculating about any universal and irreversible direction of conceptual development" in science tries "to show how the process of 'variation and selective perpetuation' helps to explain the transformations of conceptual populations; and so to restate, in a more tractable form, those questions about . . . conceptual change which . . . Kuhn left mysterious." (pp. 336–37) Indeed, Kuhn's own evolutionary metaphor from the end of *The Structure of Scientific Revolutions* is simply wrong—precisely because it tries to derive the notion of progress from that of evolution. Evolution is concerned simply with fitness for an environment not better or more developed. The "lower" can evolve from the "higher" just as readily as the "higher" from the "lower," so long as the lower fits more harmoniously into an altered ecological niche.

Leaving aside the dubious contention that the "lower" can actually evolve from the "higher" (rather than just out-reproduce the higher in some ecological niche under altered environmental conditions), there are at least two things to notice about Toulmin's analysis. First, it suggests a fifth condition—

(e) universal and necessary development—which is sometimes present or presumed in the theory of progress. The contrast between a T_1——T_2 relation that satisfies this stronger condition and one that only satisfies the previously conceptualized somewhat weaker notion of progress is not, however, crucial to the present discussion. The stronger notion can only compound the difficulties at issue. Second, Toulmin clearly presents a theory of evolutionary change that can focus questions on the issue of (d), betterness, of a specific sort, when comparing two related states of affairs in science or in technology. It is also a theory with which it is possible to conclude that neither state is better than the other—that is, neither is better adapted to its environment—in which case the relation between them can reasonably be characterized as ambiguous, although Toulmin himself does not use this term.

Once again, however, the issue here may be more problematic, or problematic in a different way, than it initially appears. Conditions (a), (b), and (c) each present their own special issues of conceptualization and confirmation. These related problems can be broached by noting that, historically, although the conception of betterness has been the focus of questioning, this conception—and hence its questioning—has taken two distinct forms: one internalist, the other externalist. The internalist view is that T_2'' brings into actuality or realizes a potency or immanent reality in T_1'. The externalist view is that T_2'' adds on (after the manner a + b) something to T_1' (containing only a). (Toulmin's theory of there either being or not being a mutually supporting fit between T and its environment can be described as a version of the externalist view, one that considers the presence of either characteristic a, fit, or b, non-fit.)

The internalist view readily applies to living organisms, and is well illustrated by the growth of most animals (with the growth of at least some plants, it is subject to question). In such cases the basic question does seem to be the exact demarcation and characteristic betterness of the mature state. What exactly defines the adult individual? In what ways is maturity better than immaturity? Perhaps a child experiences the world in a more innocent or pure way than an adult. Besides, doesn't the mature state eventuate in death? Yet it is not clear that this same internalist framework can be found in the changes undergone by species or other groups or by historical disciplines such as science—although on occasion it is presumed or argued to do so. Not to mention other problems, the original eighteenth-century theory of scientific and historical progress as eventuating in a perfected or mature stable state has clearly been overwhelmed by a history of apparently perpetual change, something not present in individuals except in a more cyclical sense.

On an externalist view, though, new problems arise focusing now not on condition (d), betterness, but on condition (a), the T_1——T_2 relationship. To some extent this T_1——T_2 relationship will include definitions of conditions (b), difference, and (c), temporal priority, along with (a), sameness. But for the internalist perspective, when focused on individuals, such conditions are more or less immediately given. Although the internalist perspective might be subject to questions of whether the chicken is really better than the egg whose potency it realizes—there are, for instance, sociobiologists who claim that the chicken is just the egg's way of creating another egg—no one would challenge the continuity between egg and chicken, or acorn and oak. The relationship is immediately given in perception; it is at root material. You can put your hand on an egg and in time you will be touching a chicken—and vice versa (at least if the chicken is a hen and you put your hand in the right place).

In scientific and technological change, however, the continuity is seldom so immediately given. This applies both to the relationship between different scientific theories (no matter how long you think Newtonian mechanics, you will never wind up thinking Einsteinian relativity) and technological objects (no matter how long you stand with your hand on a car, you will never wind up touching an airplane). It also applies to the relationships between sciences and technologies, technology and the sciences, and the societal effects of either or both. Indeed, it is no doubt precisely this lack of givenness with regard to relationship or sameness that constitutes the applicability of the externalist perspective.

Consider again the possibility of a catastrophe caused by the use of nuclear weapons, or by chlorofluorocarbon destruction of the ozone layer of the atmosphere, or by some biotechnology experiment. After such a disaster, critics commonly call into question the claims of progress made in the name of preceding technological changes on the basis of some reputed continuity between different moments in the sequence of technological changes. Different kinds of critics, of course, argue for stronger continuities and longer chains of relationships than others. At the same time, though, it is always possible for an apologist to counter by denying any inherent necessity in the relationship between the two moments $T-_1$ and T in the technological change sequence immediately preceding the disaster. All one needs to do, it could be argued, is to back up to time $T-_1$ (or re-create the history of technology up to that point) and proceed with technological development in a different direction afterward. But how is one to evaluate such claims (and counterclaims) regarding the inherent relationships between different moments in a sequence? In sequences involving human action, it is seldom easy to decide what really or necessarily follows from what. The strength and character of a continuity are problematic.

Take the example of chess. Given two good players, there are board positions in which a certain specific move will more or less automatically (provided the players are paying attention) bring in their train some sequence of further moves that will lead to checkmate. The loss of the game is then quite reasonably attributed not to the ultimate move on the part of a player, but perhaps to his penultimate or antepenultimate move, the one that set up or opened the way to ultimate downfall. There are other board positions in which one loses a game only with the ultimate move; right up until the end there is a way out. The better the players, however, the less likely this is to be the case. And because one of the players in the game of using technology—that is, nature (including human nature)—may be said to be very good indeed, it can be suggested that in this case as well complete escapes or radical changes of course will seldom be possible at the last minute.

But again, one can always respond that even with (and perhaps because of) a relatively strong continuity in a long chain of technological changes, there remains progress in the technology itself. Nuclear weapons were (assuming we are talking after a nuclear holocaust) better *qua* weapons technology than TNT, weren't they? How could a nuclear holocaust in any way alter that? Perhaps it could not, but the continuity or unity that is thus affirmed in one area (the technological) is therewith denied in another (the technology-society relationship). That is, technology, in this case weapons technology, comes to be looked upon as something almost independent of the society to which it is related. To argue for progress in technology but not its use, or for some hiatus between technological and societal development that should not be considered when judging the technology itself (as in W. F. Ogburn's "cultural lag" theory), is to postulate a discontinuity between technology and society that should challenge any apologist of progress. Indeed, it may well be that the idea of an autonomous technology lays bare precisely this aspect of the theory of technological progress.

At the very least, the judgment about progress in science or in technology will depend on what aspects of this phenomenon are allowed to count, or what reality "slices" are appealed to. What is crucial is to recognize that the factors to which one appeals are not given, except socially or historically. Such recognition conspires again to suggest that progress is not so much found in the history of science or technology as projected upon it. Immanuel Kant, a firm believer in progress, himself admitted that the question of progress is not to be answered directly from experience. In Kantian terms, progress is neither a constitutive nor even a regulative idea. Progress is an idea not derived from experience but brought to it. Moreover, such an idea is not always and necessarily brought to experience after the manner of the ideas of space and time, or even of goodness and beauty, but only historically and contingently. The character of scientific or technological change is such that its progressive features are surely, contra Laudan, at least as questionable as the rationality of science and of technology.

REFERENCES

Ellul, Jacques. "Notes on the Theme: Technical Progress Is Always Ambiguous," in *Philosophy and Technology* ed. C. Mitcham and R. Mackey (New York: Free Press, 1972), pp. 97–105.

Feyerabend, Paul. *Against Method* (New York: Schocken, 1977).

Kant, Immanuel. "An Old Question Raised Again: Is the Human Race Constantly Progressing?" (1798), in *On History,* ed. Lewis White Beck (Indianapolis: Bobbs-Merrill, 1963), pp. 137–54.

Kuhn, Thomas. *The Structure of Scientific Revolutions* (Chicago: University of Chicago Press, 1962).

Lakatos, Imre, and Alan Musgrave, eds. *Criticism and the Growth of Knowledge* (Cambridge: Cambridge University Press, 1970).

Laudan, Larry. *Progress and Its Problems: Toward a Theory of Scientific Growth* (Berkeley: University of California Press, 1977).

Mumford, Lewis. *The Myth of the Machine* 2 vols. (New York: Harcourt Brace Jovanovich, 1967 and 1970).

Ogburn, William Fielding. *Social Change with Respect to Nature and Culture* (New York: Viking, 1922).

Rescher, Nicholas. *Scientific Progress: A Philosophical Essay on the Economics of Research in Natural Science* (Pittsburgh: University of Pittsburgh Press, 1978).

Rousseau, Jean Jacques. "Discours sur les sciences et les arts" (1750).

Toulmin, Stephen. *Human Understanding,* vol. 1: *General Introduction and Part I* (Princeton: Princeton University Press, 1972).

Evolution and the Foundation of Ethics

WILLIAM B. PROVINE

INTRODUCTION

What we have learned about the evolutionary process has enormous implications for us, affecting our sense of meaning in life, our conception of our place in nature, indeed the whole foundation of ethics, including the question of whether we can make choices freely.

In a recent review I wrote: "Liberal religious leaders and theologians, who also proclaim the compatibility of religion and evolution, achieve this unlikely position by two routes. First, they retreat from traditional interpretations of God's presence in the world, some to the extent of becoming effective atheists. Second, they simply refuse to understand modern evolutionary biology and continue to believe that evolution is a purposive process."[1] An interesting response soon appeared in the form of a letter from a Presbyterian Elder, sixty-eight years old, "liberally educated at Princeton in 1939 and pretty well read in Church history and doctrine." He was writing, he said, because he fell "exactly within [my] description of those liberal religious leaders who proclaimed the compatibility of religion and evolution, and who arrived at this position by exactly the two routes [I] described."

Then came the kernel of his letter. "Where do I go from here? Is there an intellectually honest Christian evolutionist position? Can we at least salvage the historicity of the Resurrection as the German theologians Pannenberg and Hans Kung try to do? Or do we simply have to check our brains at the church-house door?" My answer to this question is that you have to check your brains at the church-house door if you take modern evolutionary biology seriously.

The implications of modern evolutionary biology are inescapable, just as the conclusion of an immense universe was inescapable when we shifted from a cozy geocentric view to the heliocentric conception of our solar system. Stated simply, evolutionary biology undermines the fundamental assumptions underlying ethical systems in almost all cultures, Western civilization in particular. The frequently made assertion that evolutionary biol-

253

ogy and the Judeo-Christian traditions are fully compatible is false. The destructive implications of evolutionary biology extend far beyond the assumptions of organized religion to a much deeper and more pervasive belief, held by the vast majority of people: that nonmechanistic organizing design or forces are some how responsible for the visible order of the physical universe, biological organisms and human moral order.

The implications of evolutionary biology for culture raise highly emotional issues. Under attack are cherished beliefs held continuously for thousands of years, going back into unrecorded history. Most people believe that a mechanistic view of life is incompatible with a decent society based upon deeply held common moral rules. This belief is untrue. The implications of evolutionary biology do demolish some basic assumptions of ethical systems, but do not destroy actual moral rules or the practice of ethics. The mechanistic Darwinian view of life is fully compatible with a deep and sensitive humanism.

In a fundamental sense, the Darwinian revolution has yet to occur in American society. In 1982, the George Gallup organization conducted a poll on belief in evolutionary biology.[2] The poll showed that 44 percent of some twenty-five hundred people polled agreed with the following statement: "God created man pretty much in his present form at one time within the last 10,000 years." By extrapolation, then, we can conclude that nearly half the people in this country are out and out creationists. For the rest, 38 percent of those polled agreed with this statement: "Man has developed over millions of years from less advanced forms of life, but God has guided this process, including man's creation"; 9 percent did not know what they believed and 9 percent accepted the scientific version of biological evolution, namely, that evolution has occurred over a long period of time and is a process that is not guided by God.

But the Gallup questionnaire did not ask a further question that could have made the results still more telling. It did not ask of that 9 percent of persons who accepted evolution without God if they believed that some other purposive force than God guided the process. I think that if Gallup had asked this question, most of those who believed that God did not guide evolution would have agreed that some other purposive force did. My conclusion is that significantly less than 9 percent of the American people believe in modern evolutionary biology. The Darwinian revolution, which is central to our understanding of evolution in nature, simply has not been assimilated or understood by the public at large. A more recent Gallup poll, conducted in 1986 for the *New York Times,* revealed that 69 percent of the Americans questioned said that God had led or guided them in making decisions, and 36 percent said that God had spoken directly to them.[3] The results of this poll suggest reasons why the public responded as it did to the earlier poll, and why the Darwinian revolution is yet to occur. Just as the abhorrence of the immense universe is understandable in the light of the earlier history of

Western civilization, so too is the opposition to the equally clear implications of Darwinism.

THE GREEK AND JUDEO-CHRISTIAN HERITAGE

Western civilization has two fundamental intellectual and religious traditions. One comes from Classical Greece, primarily through the influence of Plato and Aristotle. The other is the Judeo-Christian religious tradition. Both share a fundamental assumption: that the world and human ethics could not possibly have arisen through the sheer mechanical workings of things. Plato and Aristotle reflected Greek culture in general when they agreed upon a view of design in the world. Plato's *Timaeus* was the first of a projected trilogy in which he wished to explain the origin of order in the physical universe, and to show that the order in human life and society was based upon the universal order. The physical world, according to Plato, came into being through the action of two fundamental forces: reason and necessity. All that was harmonious and good in the world came into being through the influence of reason, personified in the *Timaeus* as the Demiurge or sometimes the lesser gods. In the absence of reason, sheer necessity yielded chaos, which was not completely without order. The mechanical interactions of physical objects guaranteed that heavy objects would settle out below lighter ones, for example. But there was no uniform passage of time, no orderly heavens, and no organized creatures.

Reason always worked with necessity in the physical world. In the final section of the *Timaeus,* Plato tells how reason cooperates with necessity to produce the human body. That the human body, or anything harmonious or good, could have come into being through sheer necessity was nonsense to Plato. Surely his view articulated an already widely held belief.

Aristotle was Plato's student for many years, but he disagreed with his teacher on many issues. Aristotle denied the existence of Plato's eternal forms and declared that the world was eternal, not created by the Demiurge or any other force. Yet the intellectual differences between Plato and Aristotle are easily exaggerated. Regarding the fundamental distinction between reason and necessity, Aristotle agreed wholeheartedly with Plato. Aristotle argued that nature works by reason for a purpose, just as a craftsman does. By necessity (efficient and material causes) alone, nothing good or organized comes about. But necessity operating under the guidance of reason (formal and final causes) yields the observable physical world. As Friedrich Solmsen has argued, the theme of nature as craftsman is fundamental to Greek thought in general.

Aristotle argued that order in nature led to ethical order in the human realm. He argued that ethical systems were in part based upon human needs, tendencies, and capacities, which vary from place to place; but all ethical

systems shared fundamental features. When Aristotle discussed human procreation in the *Generation of Animals*, he told how the particular conditions
of wombs, and the particular qualities of the sperm, yielded human babies
with observable differences, from sexual to emotional. But all babies are
human, by nature. Similarly, all human ethical and political systems share
certain fundamental generic qualities. For Aristotle, human and political and
ethical systems were created by purposive nature. This is what he meant in
his famous passage, "It is evident that the state is a creation of nature, and
that man is by nature a political animal."

For both Plato and Aristotle, human ethics was based upon order in nature.
And for both of them, order in nature resulted from purposive reason, not
from chance and necessity. The mechanical workings of things could produce
neither humans nor human ethics.

The Judeo-Christian tradition, of course, shares the same general view
concerning the roles of reason and necessity in world order and in the ethical
realm. Jehovah created and designed the heavens and earth, animals and
plants, and set moral laws for humans. Cosmic order and ethical order had
the same source. The Bible rejected the view that the world of human ethical
behavior could arise through the mere mechanical workings of things,
through chance and necessity alone. It is unsurprising that medieval philosophers could synthesize the Greek and Judeo-Christian traditions into the
powerful conceptions that have so deeply affected Western civilization.

The view that nature and humans are comprehensible only in terms of the
action of reason, as opposed to chance and necessity, has until recently
dominated Western thought. This view has also dominated in most other
cultures. Plato, Aristotle, and Aquinas were neither naive nor contradicting
available evidence in taking their positions. The general belief that reason
caused the observable order in the physical world and in ethics was pervasive
because it was obvious, emotionally satisfying, and buttressed by abundant
evidence. The evidence was that order existed in the natural world—regularities of heavenly bodies, animals and plants, and human reason—and no
compelling explanation for this widespread order could be given by mere
chance and necessity. The mechanistic conception of life has always had
adherents (in antiquity, the atomists), who demonstrate how little and for
what good reasons their views counted for little in intellectual and popular
thought.

THE SCIENTIFIC REVOLUTION OF THE SIXTEENTH AND SEVENTEENTH CENTURIES

It is often argued, by Lewis Mumford among many others, that during the
scientific revolution of the sixteenth and seventeenth centuries we started
down the awful path toward the bifurcation of science and human nature.

Galileo is particularly singled out as the key to that bifurcation, receiving a lot of attention from historians of science. After all, he had a dramatic encounter with the Catholic Church and was imprisoned in his house for the last eight years of his life, so that Galileo can serve, as he did for Andrew Dickson White, first president of Cornell University and pioneer chronicler of the conflict between science and theology, as the perfect embodiment of that conflict. I use the term "theology" advisedly. Andrew Dickson White was deeply religious. He believed the conflict was between science and theology and not fundamentally between science and religion.

Contrary to what many historians have told us, the scientific revolution did not produce a fundamental conflict between science and religion. There were, of course, many small and relatively insignificant skirmishes, of which the Galileo incident was one, but no fundamental conflict. Over against Galileo one could set the names of the deeply religious scientists, such as William Harvey, Robert Boyle, and Isaac Newton. And even the case of Galileo is equivocal. Pope John Paul II is very keen on rehabilitating Galileo as a deeply religious person and a great scientist. In the issue of *Science* for 14 March 1980, there is coverage of a presentation by Pope John Paul II to the Vatican Academy of Sciences.[4] He argued that there was no conflict between science and religion and used the case of Galileo to demonstrate his point. He said: "He who is rightfully called the founder of modern physics declared explicitly that the two truths of faith and of science can never contradict each other." And he quoted Galileo: "Holy Scripture and nature proceed equally from a divine word, the former as it were dictated by the Holy Spirit, the latter as a very faithful executor of God's orders."

The rest of Pope John Paul II's presentation follows these same lines: he makes a statement about science and religion and then reads a passage from Galileo showing that Galileo agrees with him. The final quotation from Galileo is: "Whoever labors to penetrate the secrets of reality with a humble and steady mind, is being led by the hand of God even if he remains unaware of it." Thus, if I discover anything about the natural world, God has guided my hand whether I know it or not. The profound conflict between science and religion does not begin with the rise of modern science.

CHARLES DARWIN

Charles Darwin started a real and profound conflict between religion and science, one that has only become deeper with time. What did this one man do to overthrow such powerful cultural traditions? In the first place, he was a most methodical man. An amusing, and telling, illustration of how methodical he was is provided by the careful list of pros and cons Darwin wrote down prior to his decision to marry.[5]

Under the "marry" column, Darwin noted: "children; constant companion; friend in old age; charms of music and female chitchat; good for one's health; but forced to visit and receive relations, terrible loss of time" (the negatives already begin to crop up even in the positive column). Under the "not marry" column, Darwin noted: "Choice of society and little of it; conversation with clever men at clubs; not forced to visit relatives; no expense or anxiety of children; perhaps quarreling; if many children, forced to gain one's bread—but of course it is very bad for one's health to work too much." And Darwin's conclusion? "My God, it is intolerable to think of spending one's whole life like a neuter bee, working, working and nothing after all. No, no, won't do. Imagine living all of one's days, solitarily, in smoky, dirty London house. Only picture to yourself a nice soft wife on a sofa with a good fire, and books and music perhaps. . . . Marry, marry, marry. Q.E.D."

This was a very methodical man. Michael Ghiselin claims that Darwin's hypotheticodeductive method was the key to his success as an evolutionary biologist, and it is certainly a part of the explanation.[6] Also important were Darwin's experiences on the voyage of the *HMS Beagle* and his intense desire to generalize about his meticulous observations.

His truly revolutionary achievement was to invent an explanation for the primary mechanism of evolution, natural selection, which ran against the grain of both the Greek and Judeo-Christian sources of Western civilzation. Indeed, I would cast a net that is wider even than Western civilization. I suspect that as Western culture did, most cultures have evolved views of purposive nature underlying the foundations of human ethics, so that the net that I am casting here also extends over all those cultures. It applies equally to any cultural group that has a purposive view of nature, of how nature works and of any relation whatsoever between those purposes and purposes in the realm of human values and ethics.

Darwin's idea of evolution by means of natural selection made evolutionary change the result of a process—I will not call natural selection a mechanism, because it is no separable mechanism of any sort whatsoever—that is itself the result of many very complex materialistic processes that occur in natural populations. Natural selection is simply Darwin's name for what happens when these materialistic forces operate on natural populations and cause some individuals or segments of the population to increase more rapidly than other segments. It is a totally purposeless process, one that is completely opportunistic, with no sense within it of a larger design of any sort. Yet, as Darwin pointed out, natural selection was capable of generating the most exquisite adaptations in organisms, precisely the kind of adaptations in which proponents of natural theology had for centuries found evidence for the hand of God.

Darwin was the first evolutionist to convince most biologists, indeed most educated people generally, that evolution had in fact occurred. He did not, however, convince many people that evolution had occurred primarily by his process of natural selection, and I think that there are two reasons for this failure.

One reason was that he had no direct evidence of natural selection. The most cursory examination of *On the Origin of Species* reveals that in the first four chapters, where Darwin introduces the idea of natural selection, his only examples are imaginary. At the end of the fourth chapter he asks the reader to judge natural selection as an explanation of evolution on the basis of the tenor of the evidence to be presented later in the book. Darwin offers in the rest of the book a wealth of evidence about geographical distribution, the geological record, classification, and embryology; but the best that he could do for natural selection was to chant it over the evidence presented at the end of each section. "Look," he seems to be saying, "everything that I have said in these chapters is explainable on the basis of evolution by natural selection."

Of course, everything he had writtten was also explainable by other evolutionary factors, as, for example, by the inheritance of acquired characteristics (which Darwin accepted), or by some kind of purposive or directive process. The evidence of paleontology can be fitted to virtually any evolutionary process that one can imagine.

The second reason for his failure to convince others of the power of natural selection was even more fundamental. People understood full well that natural selection undercut the deepest commitments of Western civilization; this implication they found disgusting. The result of accepting Darwin's idea of evolution by descent but rejecting his theory of natural selection was, in the late nineteenth century, a wild profusion of theories of evolution and of heredity, literally hundreds of them, the vast majority involving some kind of purposive mechanism, either in the hereditary process or in the evolutionary process. In this way the originators of the theories hoped to save the cultural traditions while accepting the scientific idea of evolution.

EVOLUTIONARY THEORY IN THE TWENTIETH CENTURY: THE EVOLUTIONARY CONSTRICTION

In the twentieth century, an amazing thing happens to all but a few of the theories of heredity and the theories of evolution that were ubiquitous in 1900. They were killed by twentieth-century biology. Mendelism killed all the theories of heredity. In 1903, the great French biologist Yves Delage published the second edition (first edition 1895) of his huge book, *L'Hérédité et les grandes problèmes de la biologie générale,* which was a complete compendium of the theories of heredity then considered important. The 1903 edition does not mention Mendel. Mendelian heredity was rediscovered in 1900 and within fifteen years (mostly within ten) it had killed all of the previous theories of heredity that Delages had discussed so carefully. Considered a very important book at the time of original publication in 1895, the second edition sank into immediate and complete oblivion. By 1915, all that remained was Mendelism and, as a minor strain, the neo-Lamarckian theory of inheritance of acquired characteristics.

The same sort of thing happened to the many evolutionary theories that avoided natural selection and attempted to infuse purpose into the evolutionary process. Some of these theories persisted into the 1920s and 1930s, a few even into the 1940s. All that remained were killed off during the period of the "Evolutionary Synthesis," in the 1930s and 1940s. In 1980, Ernst Mayr and I edited a book with the title *The Evolutionary Synthesis*,[7] but now I think that title was misleading because I no longer think that there was a real synthesis of the disparate parts of evolutionary biology—systematics (the study of classifications), paleontology, embryology, psychology, and so forth—with Darwinian natural selection and Mendelian genetics into a grand unified theory of evolution in nature. That is not what happened, in spite of Ernst Mayr and I giving that impression in 1980. Instead, I now believe that what really happened in evolutionary biology during the 1930s and 1940s was that the great profusion of evolutionary theories prior to the 1930s were killed off by evolutionary biologists who argued that there were many fewer important variables in the evolutionary process than scientists had earlier thought.

Beginning in the late 1910s, the theoretical population geneticists R. A. Fisher, J. B. S. Haldane, and Sewall Wright claimed that the evolutionary process could be modeled quantitatively and they soon provided such models. The prestige of the quantatitive modeling, the application of the new genetics to systematics by Mayr and to paleontology by Simpson, combined with the dying out of the older generation of biologists who believed in purposive mechanisms in evolution, yielded a new conception of evolution in nature. This new conception was not, however, unified and synthesized. After all, biologists have been embroiled in almost constant controversy in evolutionary constriction" instead of the "evolutionary synthesis." This term conceptualization did do was to kill off the other conceptions so that now we have left a tiny subset of variables that determines the evolutionary process.

All of the variables in modern evolutionary biology are in that subset of variables that the authors of the new conception focused upon, so I now call what happened to evolutionary biology in the 1930s and 1940s the "evolutionary constriction" instead of the "evolutionary synnthesis." This term seems to me to be a more accurate description of what occurred.

One of the effects of the evolutionary constriction, in ridding evolutionary biology of all purposive theories of evolution—a step that virtually all evolutionary biologists working today agree was a progressive development—was to make the conflict between science and religion inescapable. Before the early 1930s, it was entirely possible to evade the conflict between evolution and religion. For example, who was the dean of American evolutionary biologists in 1925? This same evolutionist spoke for evolutionists at the

Scopes "Monkey" trial in Dayton, Tennessee, and wrote a little book about the trial entitled *The Earth Speaks to Bryan*. It was Henry Fairfield Osborn of the American Museum of Natural History. Osborn's own theory of evolution was called "Aristogenesis." It would be a challenge to find reference to it in a modern textbook because Aristogenesis does not not exist any more. It was a purposive theory of evolution, reflecting Osborn's deep Christian religiosity. He saw no conflict at all between science and religion. Indeed in *The Earth Speaks to Bryan* he told William Jennings Bryan, the creationist lawyer in the Dayton trial:

> If Mr. Bryan, with an open heart and mind, would drop all his books and all the disputations among the doctors and study first-hand the simple archives of Nature, all his doubts would disappear; he would not lose his religion; *he would become an evolutionist*.[8]

What modern evolutionary biology has done to us, however, is very different from what Osborn believed the study of nature would do for Bryan. It tells us (and I would argue that the same message flows in from physics, chemistry, molecular biology, astrophysics, and indeed from all of modern science) that there is in nature no detectable purposive force of any kind. Everything proceeds purely by materialistic and mechanistic processes of causation or through purely stochastic processes. There may be quantum mechanical indeterminacy and it may be absolutely fundamental, allowing no hope of going beyond a probabilistic description of nature at the quantum level, but all this means is the ultimacy of chance, although it should always be remembered that chance processes at the quantum level become the determinism of statistical laws at the level of many quanta, where all biology rests. All that science reveals to us is chance and necessity, as Jacques Monod argued.[9] The problem with Monod's argument, however, is that he only saw the destructive implications of that revelation, not the constructive ones.

IMPLICATIONS OF MODERN SCIENCE

First, modern science directly implies that the world is organized strictly in accordance with mechanistic principles. There are no purposive principles whatsoever in nature. There are no gods and no designing forces that are rationally detectable. The frequently made assertion that modern biology and the assumptions of the Judeo-Christian tradition are fully compatible is false.

Second, modern science directly implies that there are no inherent moral or ethical laws, no absolute guiding principles for human society.

Third, human beings are marvelously complex machines. The individual human becomes an ethical person by means of two primary mechanisms:

heredity interacting with environmental influences. That is all there is. I think that E. O. Wilson and some of the other sociobiologists who work on humans have exaggerated the amount that heredity determines human moral behavior, but that matter is by no means settled yet.

Fourth, we must conclude that when we die, we die and that is the end of us. That is what modern science tells us. Bertrand Russell put it very nicely. He said: "I believe that when I die I shall rot, and nothing of my ego will survive. I am not young, and I love life. But I should scorn to shiver with terror at the thought of annihilation. Happiness is nonetheless true happiness because it must come to an end, nor do thought and love lose their value because they are not everlasting. . . . Even if the open windows of science at first make us shiver after the cozy indoor warmth of humanizing myths, in the end the fresh air brings vigor and the great spaces have a splendor of their own."[10]

I think the last part of Russell's quote is very poetic, smacking of hopeful nonsense, but the first part is right on. When I die I shall rot and that is the end of me. There is no hope of life everlasting. Even card-catalog immortality (a big enough section of a library catalog to be noticed by future scholars) is ephemeral at best.

Finally, free will as it is traditionally conceived, the freedom to make uncoerced and unpredictable choices among alternative possible courses of action, simply does not exist. That assertion is often more difficult to accept than the assertion that there is no God, and for good reason. There are many ethical traditions that have been built upon atheism, but most of them are founded upon some secular notion of humans having the free will to make moral choices and thus bear true moral responsibility. What modern science tells us, however, is that human beings are very complex machines. There is no way that the evolutionary process as currently conceived can produce a being that is truly free to make choices.

Ethical philosophers generally disagree strongly with my argument that free will does not exist. One group of them, the majority, argues that there is no contradiction whatever between determinism and free will. This group argues that free will is like the freedom of a weathervane to turn freely in the wind, or like free wheeling in a car. This argument reminds me of the fishmonger who shouted for hours "fresh fish for sale, fresh fish for sale." Toward the end of the day, a customer, attracted by the nicely shouted advertisement, discovered that the fish for sale smelled bad, and told the fishmonger that the fish were not fresh as he was shouting. The fishmonger replied, "Fresh fish is just their name. I never said the fish were fresh." Those who shout the compatibility of determinism and free will are merely calling human decision making "free will." The direction a weathervane points is determined by the direction of the wind and is in no way free. Free wheeling in a car simply means that the drive train is disconnected from the differential and the speed of the car is determined by its mass, air resistance, resistance in the mechanisms of the car, and the force of gravity on whatever incline the car happens to be.

Other more honest ethical philosophers, who actually believe in free will, argue for an incompatibilist position on determinism and free will. These philosophers think that in nature there is both determinism and indeterminism, as in quantum mechanical indeterminism. They argue with good reason that determinism and free will are incompatible. But they also believe in free will. This is a very difficult position to take because none of them can figure out a direct way that any of the indeterminism that we know about in nature could possibly, much less plausibly, produce human free will. So, like the philosopher Peter Van Inwagen at Syracuse University,[11] they use a backhanded logical argument that proceeds like this: If we had no free will, then there could be no moral responsibility. But some people are morally responsible. Therefore, some persons have free will, which, because it cannot come from determinism, must emanate from the indeterminism in nature.

This argument is appealing but demonstrably wrong. The weak assumption is that without free will there could be no moral responsibility. The only way that biological organisms can exhibit moral responsibility is for them to be trained to have it. We humans spend a vast amount of time trying (not very efficiently, I think) to make our children into morally responsible individuals. Moral responsibility is wholly compatible with no free will, indeed is impossible with it.

My conclusion about these two warring schools of ethical philosophers is that they are both right. Those who are compatibilists about free will and determinism are right that the other group cannot get free will from indeterminism in nature. The second group is correct that the first cannot get free will from determinism. The reasonable conclusion, the one that is consonant with modern science, is that there is no free will. This view is advocated by only a tiny minority of brave ethical philosophers, all of whom deserve our sympathy for enduring the barbs of their colleagues.

ATTEMPTS TO ESCAPE THE INEVITABLE

The squirming and squiggling to escape the implications of modern science are something to behold, and just as interesting among scientists as among federal judges, humanists, or liberal theologians who distrust science. I will begin with the United States National Academy of Sciences, one of the most prestigious groups of scientists in the world. A recent booklet on science and creationism published by the academy opened with a preface signed by the president, Frank Press. He stated: "It is false, however, to think that the theory of evolution represents an irreconcilable conflict between science and religion. A great many religious leaders and scientists accept evolution on scientific grounds without relinquishing their belief in religious principles."[12]

There is no conflict between science and religion according to the National Academy of Sciences.

At the recent Arkansas creationism trial (*McLean* v. *Arkansas Board of Education,* 1982), Judge William R. Overton said in his decision that we must follow the version of the Establishment Clause of the first amendment to the Constitution found in *Lemon* v. *Kurtzman* (1973), which stated a three part requirement. First, a statute must have a secular legislative purpose. Second, the principal or primary effect of the statute must be one that neither advances nor inhibits religion. And third, the statute must not foster an excessive government entanglement with religion. Judge Overton ruled that the Arkansas statute mandating the teaching of creation "science" failed on all three counts. His decision was applauded by scientists, theologians, religious leaders, sociologists, philosophers, and jurists.

At the trial and later, creationists argued that if the principal or primary effect of a statute must be one that neither advances nor inhibits religion, then it must be unconstitutional to teach evolution. Evolutionism contradicts our religion and thus inhibits our religion. Judge Overton's response at the trial was to say that evolution was in no way antithetical to religion: "Evolution does not presuppose the absence of a creator or God. The plain inference conveyed [by the creationists] is erroneous." And he adds a footnote: "The idea that belief in a Creator and the scientific theory of evolution are mutually exclusive is a false premise and is offensive to the religious views of many. Dr. Francisco Ayala, a geneticist of considerable renown and a former Catholic priest who has the equivalent of a Ph.D. in theology, pointed out [to this court] that many working scientists who subscribed to the theory of evolution are devoutly religious."[13]

We have now seen the same argument twice, once from the National Academy of Sciences and once from a federal judge. The argument is that some evolutionary biologists are also religious, so there cannot be conflict between modern science and religion. Instances of the invocation of this argument are found everywhere. I have two objections to the argument. The first is that very few truly religious evolutionary biologists remain. Most are atheists, and many have been driven there by their understanding of the evolutionary process and other science. Second, it is no argument for the compatibility of science and religion that some few evolutionary biologists see no conflict. To my mind, they are either obtuse or compartmentalized in their thinking, or are effective atheists without realizing it. According to the unlikely coalition of scientists, jurists, educators, theologians, and religious leaders, the conflict between science and religion exists only in the naive, literalist minds of the creationists. Scientists testify that creation science is stupid and so is the literalist religion of the creationists. The federal courts and now even the Supreme Court (in *Edwards* v. *Aguillard,* 1987), say the same things. So do the religious leaders. Nearly half of the American public is being called stupid or at best misguided. I agree that evolution has occurred and that creationists are wrong about that. But they are right that there is a conflict between science and religion, and not only their religions.

There is a real conflict. The problem is fundamental and goes much deeper than modern liberal theologians, religious leaders, and scientists are willing to admit. Most contemporary scientists, the majority of them by far, are atheists or something very close to that. And among evolutionary biologists, I would challenge the reader to name the prominent scientists who are "devoutly religious." I am skeptical that one could get beyond the fingers of one hand. Indeed, I would be interested to learn of a single one.

I believe that there is a lot of intellectual dishonesty on this issue. Consider the following fantasy: the National Academy of Sciences publishes a position paper on science and religion stating that modern science leads directly to atheism. What would happen to its funding? To any federal funding of science? Every member of the Congress of the United States of America, even the two current members who are unaffiliated with any organized religion, professes to be deeply religious. I suspect that scientific leaders tread very warily on the issue of the religious implications of science for fear of jeopardizing the funding for scientific research. And I think that many scientists feel some sympathy with the need for moral education and recognize the role that religion plays in this endeavor. These rationalizations are politic but intellectually dishonest. How many religious leaders are atheists or close to it? How many would be honest about any conflict they might see between religion and science?

Many theologians have reacted to the rise of modern science by retreating from traditional conceptions of God and its presence in the world, calling this a more sophisticated view. God used to be all around us earlier in our cultural history—used to do miracles, to guide its people. People could detect God's presence all the time; but times have changed. God is more remote today. In fact, one cannot rationally discover anything that God does in the world anymore. A widespread theological view now exists saying that God started off the world, props it up, and works through laws of nature, very subtly, so subtly that such action is indetectable. But that kind of God is effectively no different to my mind than atheism. To anyone who adopts this view I say, "Great, we're in the same camp; now where do we get our morals if the universe just goes grinding on as it does?" This kind of God does nothing outside of the laws of nature, gives us no immortality, no foundation for morals, or any of the things that we want from a God and from religion.

MEANING IN LIFE

A friend of mine, Jack Haught, who teaches theology at Georgetown University in Washington, D.C., has written several books arguing for the compatibility of religion and science.[14] He is very keen on having cosmic meaning in life. He has a logical argument that is reminiscent of the one I presented earlier about free will and moral responsibility, and it runs like this: Without cosmic meaning, our lives would be meaningless; but some humans have very meaningful lives; thus, there must be cosmic meaning. I can see no cosmic or ultimate meaning in human life. The universe cares nothing for us and will either continue to expand and cool, leading to an extermination of all living creatures in the universe, or the universe will cease to expand and will begin collapsing, which will result in everything crashing together in an unbelievably small space, thus obliterating all life. Humans are as nothing in the evolutionary process on earth, and only a few individuals are remembered for as many as ten generations. There is no ultimate meaning for humans.

But certainly humans can lead meaningful lives. My own life is filled with meaning. I am married to a talented and beautiful woman, have two great sons, live on a beautiful farm with lots of old but good farm machinery, teach at a fine university where the students are excellent, and have many wonderful friends. But I will die and soon be forgotten. Jack Haught will have a tough time convincing me that my life is meaningless just because there is no cosmic meaning for it, or that all the meaning that makes my life worthwhile is really cosmic meaning.

FOUNDATION FOR ETHICS

We need to recognize what modern science has done to us and try to come to terms with its implications for the foundation of morality. A growing number of young persons no longer believe in a god who lays down moral laws for them and punishes them if they disobey it. Many of them conclude that they can therefore behave in any way they can get away with. At the same time, we are surrounded by terrible moral behavior on the part of adults. The only approach that I can hope will help the growing crisis in morality is through understanding that we humans are just complex machines without free will that have been poorly programmed for moral behavior. We are abysmally ignorant of the moral development of small children and are repulsed by the notion that they will grow up to be the moral persons we program them to be. B. F. Skinner long ago understood that we could really control the moral development of small children. In a general sense he is surely correct no matter how repulsive is the truth. We must learn more about how to make little children more surely into moral beings and help parents like me and my wife to better program our children. If this seems repellent, is it any better to do the same thing anyway, but do a much worse job of it because we refuse to think about what we are really doing?

We also must encourage people, especially young people, to think rationally about moral behavior. They should not behave well because they will be punished by a purposive moral force if they do not. They should behave well because if they do not they are behaving foolishly in terms of their own interests. If you do not behave altruistically, and you reinforce other people's nonaltruistic behavior, then you are helping to generate a very poor society in which to live. It is in our own best interest to behave altruistically. We can get more out of life with good and supportive friends. Indeed, because there is no almighty friend, the only ones we have are our human friends, and perhaps even animal and plant friends.

I grew up in Nashville, Tennessee, and frequently went to the Grand Ole Opry. I remember particularly one gospel song with this line in the chorus: "If you don't love your neighbor, then you don't love God." I think that what modern science tells us is: "If you don't love your neighbor, then you're just plain stupid!"

NOTES

1. Provine, W. B., "Trial and error: The American controversy over creation and evolution," *Academe* 73 (1987), p. 52.

2. Reported in *The New York Times,* 29 August 1982.

3. Reported in *The New York Times,* 10 December 1986.

4. "Address of Pope John Paul II," *Science* 207 (1980), 1165–67.

5. As reported in de Beer, G., *Charles Darwin: A Scientific Biography,* (New York: Anchor, 1965), p. 110.

6. Ghiselin, M., *The Triumph of the Darwinian Method,* (Berkeley: The University of California Press, 1969).

7. Mayr, E. and W. B. Provine, eds., *The Evolutionary Synthesis,* (Cambridge: Harvard University Press, 1980).

8. Osborn, H. F., *The Earth Speaks to Bryan,* (New York: Charles Scribner's Sons, 1925), pp. 20–21.

9. Monod, J., *Chance and Necessity,* (New York: Knopf, 1971).

10. As cited in *The Encyclopedia of Philosophy,* (New York: Macmillan, 1967) vol. I, p. 188.

11. Van Inwagen, P., *An Essay on Free Will,* (New York: Oxford University Press, 1982).

12. *Science and Creationism: A View from the National Academy of Sciences,* (Washington, D.C.: National Academy Press, 1984), pp. 5–6.

13. "Creationism in the schools: The decision in McLean versus the Arkansas Board of Education," *Science* 215 (1982), pp. 934–43.

14. See especially Haught, J. F., *The Cosmic Adventure: Science, Religion, and the Quest for Purpose,* (New York: The Paulist Press, 1984).

Perils of Progress Talk: Some Historical Considerations

JOHN M. STAUDENMAIER, S.J.

INTRODUCTION

By most measures, 1933 was the worst year in America's long economic history. Unemployment, in the fifth year of a savage depression, reached its highest level, 25.2 percent.[1] Bread lines and homeless wanderers provided the body politic with vivid images of shattering failure in the body economic. Still, visitors to Chicago's International Exposition that year saw no sign of social pain. The dazzling spectacle (boasting "more colored lights . . . than in any equal area or even any city of the world") was dedicated to a "Century of Progress." In the official Guidebook's effusive prose:

> The dawn of an unprecedented era of discovery, invention, and development of things to effect the comfort, convenience, and welfare of mankind . . . An epic theme! You grasp its stupendous stature only when you stop to contemplate the wonders which this century has wrought.[2]

A more blatant example of Depression incongruities would be hard to imagine: overwhelming human suffering juxtaposed with unabashed huckstering. It is tempting to explain the Exposition's upbeat theme as simply another example of the popular escapism manifest in the spate of entertainment movies so popular during the era. Images of prosperity provide some slim comfort in grim times. Such an explanation ultimately falls short, however, because it overlooks ominous undertones found in the fair's iconography and guidebook rhetoric. Three examples suffice. "Science" and "Industry," the second term a conflation of technology and business, were represented by two forty-foot, aluminum coated figures frowning down upon entrants to the Hall of Administration. Even more striking, visitors to the Hall of Science were met in the foyer by the "Fountain of Science" with Louise Lentz Woodruff's three-piece sculpture, "Science Advancing Mankind." Two life-size figures, male and female, faced forward with arms uplifted. Both were dwarfed by the massive figure of a robot twice their size. In the words of Lenox Lohr, general manager of the exposition, the robot typified "the exactitude, force and onward movement of science, with its hands at the backs of the figures of a man and a woman, urging them on to the fuller life."[3] Finally, the sculpture's iconographic message was aptly reinforced by the Guidebook's stunning, bold-faced thematic motto: "SCIENCE FINDS, INDUSTRY APPLIES, MAN CONFORMS."[4]

The Exposition, in short, celebrated inevitable scientific-technological progress while reducing the human role in the process to conformity. How can we explain the paradox inherent in such visual and rhetorical imagery? How have Science, Technology, and Business, the three prime agents in the West's claim of human triumph over life's ancient constraints, come to be linked with passive conformism? In what follows, I will argue that the connection is much more than an oddity unique to Chicago and 1933. Rather, it lies at the core of one form of progress ideology that I shall call "autonomous progressive determinism." My purpose here is to explore some of its origins and to critique its role in our contemporary discourse. Put briefly, the ideological position I am concerned with argues the following six points: that since the 17th-century "Scientific Revolution," humanity has had access to a radically new form of knowledge, sometimes called the scientific method, sometimes called the controlled-variable experimental method; that the scientific method yields value-free truth, which liberates humanity from the bias and superstitution of prior forms of knowledge; that the 18th-century "Industrial Revolution" created a radically new form of technology whose methodological power depends on its primary characteristic, namely, the application of science; that the two methods demand freedom from the influence of all outside controls; that the emergence of capitalism during the same Industrial Revolution marks the beginning of an economic method (the invisible hand of the free market) similarly free from the exterior constraints of tradition and bias; and that these three new forces, taken together, operate as an inevitable, autonomous force called "Progress," which continually advances humanity's conquest of nature.

It has become abundantly clear that I am not enamored of this rhetoric of progressive determinism. Before turning to my specific critiques of the position, however, a disclaimer is in order. I do not mean to argue here that science, technology, or business, terms used for a wide variety of human endeavors, are simply evil or good. Nor do I want to suggest that various scientific and technological traditions have not been cumulative, that earlier science or technology has not influenced what has followed. My position is, rather, that the ideology of progress as an inevitable deterministic process has operated as a quasi-religious belief in the West and that, to some extent, all citizens of the West carry fragments of it embedded in our language and attitudes. Few, if any, still defend this pure position. Why then take the trouble of detailed discussion?

Progress talk remains influential today as a language that normally operates on the semiconscious level of uncritical assumptions. It fosters some modes of discourse while rendering others less accessible and impoverishing our imaginations in the process. Inherited language influences us all; we remain citizens of our traditional culture. Paying attention to the pure position alerts us to our various ways of buying into the ideology both in our linguistic habits and our affective responses to scientific, technological, and business issues.[5]

I will begin by trying to sort out several conceptual strands that frequently appear in discussions of progress ideology, paying particular attention to the United State's context and to several remarkable shifts in the popularly accepted meaning of progress between the Revolution and the present. I will then discuss its roots in European philosophical discourse and analyze what I take to be its pivotal assumption, the claim that scientific, technological, and business methods must operate independently of their contexts. After criticizing the rhetoric of autonomous progress on historical and logical grounds, I will suggest two destructive consequences of this linguistic heritage; the legitimation of Western colonial violence and the inculcation of political passivity among citizens of the West. Finally, and briefly, I will refer to my own alternative model for a language that interprets technological change without succumbing to the seduction of progress talk.

PROGRESS TALK IN UNITED STATES DISCOURSE

"Progress," as an explanatory model for history, has proved a slippery concept. Several thoughtful essays about its place in the United States context provide a helpful starting point. Leo Marx and Merritt Roe Smith have called attention to a significant shift in the concept of progress that had dramatically affected its role in shaping the nation's self-understanding.[6] Marx begins by describing the earlier "Enlightenment" approach:

> The concept of history that won favor at the time [18th century] turns on the idea of "progress": the belief that history is the record of a steady, continuous, cumulative, and [in its most extreme form] a somehow preordained improvement in the material, social, cultural—indeed, the overall—conditions of life, and that its driving force is the expansion of human knowledge and power achieved by science and technology.

A page later, he stresses its distinction from what would follow.

> What made the seminal form of the Enlightenment commitment to progress so different [from the more recent concept], indeed, was *an accompanying vision of radical political liberation.*[7] (My emphasis)

Marx and Smith see Thomas Jefferson's insistence on the equal importance of moral and material improvement as typical of the Enlightenment vision.

Progress, for Jefferson and like minded 18th-century thinkers, depends on continual human efforts toward personal and political liberation, toward the formation of virtue in the citizenry of the new Republic. To be sure, progress also flows from the scientific and technical endeavor, but care must be taken not to upset the balance between these two elements.[8]

Sometime after the turn of the 19th century, however, a marked shift in the rhetoric of progress becomes evident.

> Slowly but perceptibly, the belief in progress began to shift away from the moral and spiritual anchors of the revolutionary era toward *a more utilitarian and hard-headed, business-oriented emphasis on profit, order, and prosperity.*[9] (My emphasis)

Thus, Commissioner of Patents Thomas Ewbanks stresses the human capacity as manufacturer while giving short shrift to the Jeffersonian political ideal.

> While most persons think not, and care not, what the prominent character of the planet is, many view it in aspects congenial to themselves: *a theatre for politicians,* a battle-field for warriors, a court for lawyers, a loungingplace for people of fashion and leisure, etc. . . . while the Indian believes it was made for nothing else than hunting game in. With these, and all surface dreamers, its vast underground treasures are not thought of. The inorganic world, its forces, principles, and processes, are to them as if they were not. It is *only as a Factory,* a General Factory, that the whole materials and influences of the earth are to be brought into play.[10] (My emphasis)

Ewbanks is only one example of the rising rhetorical tide celebrating 19th-century America's technological triumphs that would culminate, as Marx puts it, in "a thoroughly technocratic idea of progress."[11]

Helpful as these essays are for nuancing the changing character of progress talk, one could read them and overlook what seems to me to be the crucial characteristic of the 20th-century version, the radical human passivity implicit, (or, as in one Chicago exposition's rhetoric, explicit) in the concept of technology as an autonomous force at work in history. It would be possible to read Smith's "hard-headed business oriented emphasis on profit, order, and prosperity" as a celebration of man precisely as active, not passive, as conqueror of nature and creator of wealth. Thomas Ewbank's vision of man the manufacturer was hardly unique.[12] How, then, might we explain the shift from man as active and aggressive to the passive and conforming man of the Chicago Exposition?

In their provocative essay, "The Mythos of the Electronic Revolution," Carey and Quirk propose an American shift from the encounter with nature as the Jeffersonian source of republican virtue, through a period of disillusionment with the social violence of steam-powered factory mechanization, to a new democratic ideal, which they name "the electronic sublime." Citing a host of social theorists, from Patrick Geddes and the early Lewis Mumford to David Lilienthal and Marshall McLuhan, they point to a popular electronic dream promising "freedom, decentralization, ecological harmony, and democratic community"—in short, the very combination of political liberty and scientific-technological progress seen in Jefferson. Citing the same array of authors, however, Carey and Quirk underscore the ironic electronic reality that has subverted the enlightenment dream. Thus, for McLuhan, "electromagnetic technology requires *utter human docility and quiescence*"[13] (my emphasis).

Helpful as it is, Carey and Quirk's reading of these contradictory strands in twentieth-century electronic ideology places too much emphasis on a single source for the shift toward passivity. Electronic systems, though clearly important in their own right, are better understood as one among a host of social and technological developments that, beginning as early as 1890, foster a vision of human passivity in the face of scientific and technological progress.

Gilded-age America, for all its enchantment with protean technological triumphs, was racked with urban violence. Bloody confrontations between workers and management police—from the Molly Maguires of 1876 through the railway strikes of 1894—were covered in lurid, and generally antiworker, detail in the national media.[14] Millions of immigrants from the hitherto unfamiliar countries of eastern and southern Europe flooded the nation with what seemed to many to be hordes of frighteningly un-American strangers.[15] At the same time, American inventiveness in the area of standardized system design (rails, telegraph, telephone, and electric utilities in particular) marked the rise to cultural dominance of a technological style based on solving problems through complex standardized systems and the gradual replacement of an earlier style that depended on the more political process of negotiation among technological practitioners, workers, and local citizenry.

Examples abound. Consider, for instance, the change in technological style both embodied and fostered by the railroad. In its first half-century of existence, the typical railroad evolved from the turnpike model—state-owned and state-supported roadbed—to a private, centrally owned enterprise that included roadbed and most system components. In the process, the relationship of railroads to their surroundings changed dramatically. Historian J. L. Larson provides an example by contrasting the design of grain-shipping facilities in St. Louis and Chicago in 1860. The St. Louis design demanded bagging the grain, loading it onto train cars, off-loading it at the outer edge of town where the tracks ended, teamstering it across the city and loading it again onto river boats. The Chicago design permitted bulk loading onto grain cars because the company-owned track ran all the way to the docks, where it was off-loaded onto grain boats. Larson concludes his description with the following provocative sentence:

If the Chicago system was a model of integration, speed, and efficiency, the St. Louis market preserved the integrity of each man's transaction and employed a host of small entrepreneurs at every turn—real virtues in ante-bellum America.[16]

The St. Louis setup required negotiation as a part of the shipping process, whereas the more complex and capital-intensive Chicago design achieved greater efficiency and permitted railroad management to ship grain without needing to negotiate with that "host of small entrepreneurs at every turn."

In like fashion, the American factory system transformed the relationship between manager and worker from the sometimes respectful and sometimes tumultuous interaction of the early American small shop to heavy-handed enforcement of work rules coupled with the deskilling of workers through the use of increasingly automated machines.[17]

The kinds of negotiation—highly skilled workers with owners, or local businessmen with national rail lines—preempted by the newly rationalized systems were a messy, unpredictable affair, often calling for high levels of political skill to achieve technical results. The new standardized systems, then, modeled the ability to exert control of unruly variables through elegant system design for a nation beset with troublesome signs of impending social chaos. It is not surprising that those Americans who sought a return to social order would look to the same scientific and technological prowess for an answer. Thus, in 1898, leading sociologist Edward Alsworth Ross urged "the right persons" (i.e., social scientists) to undertake "the study of moral influences. . .in the right spirit as a basis for *the scientific control of the individual*"[18] (my emphasis). Ross warned against revealing these scientific secrets for "to betray the secrets of social ascendancy is to forearm the individual in his struggle with society."

Seen from this perspective, Progressive-era reformers take on a new significance. Their commitment to "rationality" and "science" as the chief means for attaining the new social order marks the definitive end of an earlier era's assumption that political negotiation lay at the heart of the American dream. Robert Bellah and his coauthors of the recent *Habits of the Heart* describe the change in striking terms:

> This desire for a more "rational" politics, standing above interest but based on *expertise rather than wisdom and virtue,* moved American political discourse away from concern with *justice,* with its civic republican echoes, toward a focus on *progress*—progress defined primarily as material abundance.[19] (My emphasis)

As Marx and Smith have indicated, the movement to define progress in terms of material abundance begins well before the Progressive Era. What is new, then, is the gradual abandonment of trust in the interactive creativity of human beings as the source of political cohesion and national self-identity. At the turn of the century, that confidence begins to be replaced by trust, not so much in the "scientific experts," as in the "Science" they represent.

On another front, the Ford labor reforms of 1914 countered the previous year's astonishing 370 percent turnover among workers on the company's nearly complete, moving assembly line with a mixture of in-factory spies and home-visiting inspectors aimed at producing stable, conforming, and Americanized workers for the company.[20] Ford's mix of enforcement and paternalistic betterment programs, perhaps the most successful of prewar attempts at "Scientific Management," proved to be an early version of postwar control tactics seen in the Red Scare of 1919–20, tear gas in 1923, industrial psychology, and welfare capitalism plans throughout the 1920s.[21]

Perhaps most striking of all, a dramatic shift in the character of advertising took place in precisely the same time frame. Nineteenth-century advertisements often took the form of a dialogue that assumed a basic equality between advertiser and reader wherein sales were thought to result from a rational dialogue about product qualities. World War I marked the rise to prominence of a style focusing, not on product qualities, but on what the product could do for the consumer. Advertising rhetoric was aimed at a consumer who was presumed to be irrational and basically inept. Roland Marchand describes the basic mentality:

> In viewing the urban masses, advertisers associated consumer lethargy as much with weak-kneed conformity as with cultural backwardness. . . . Emotional appeals succeeded because only by seeking this lowest common human denominator could the advertiser shake the masses from their lethargy without taxing their limited intelligence.[22]

In summary, the shift in progress rhetoric from a vision of man as aggressive conqueror of nature and creator of wealth to the present century's passive conformist results not from the symbolic power of a single technology, as Carey and Quirk suggest, but from an obsession with social, scientific, and technological control pervading level after level of national life.[23] Just as inventors and entrepreneurs embodied the advance of progress in standardized system designs, so advertisers, social psychologists, and corporate managers worked to standardize the interior motivation of workers and consumers in their quest for a stable social order. As the 1933 Chicago Exposition indicates, not even the Depression's massive economic collapse would significantly change the core ideological premise of the new vision of progress. "Progress" had come a long way from the founding fathers' Enlightenment vision with its confidence in the capacity of human beings to transform nature for their benefit while negotiating a liberating political order. By 1930 "Progress" implied radical human passivity in the face of the twin gods, "Science" and "Technology."[24]

AUTONOMOUS PROGRESS AND THE SPLIT OF METHOD FROM CONTEXT:

From Descartes to Capitalism: The Philosophical Tradition

Still, this remarkable transition from Enlightenment creativity to 20th-century passivity may not be as surprising as it seems. The ideal of progress traces its philosophical origins at least back to Descartes. Latent in that tradition we can see a central premise that, while originally masked by Enlightenment optimism about the scientific, technological, and political endeavors, provides another strand in the explanation of how Americans came to see Science and Technology as autonomous social forces and themselves as conformists.

Descartes posited a radical disjunction of valid method from human context, a supposition that may well prove the most important single assumption governing the West's commitment to progress as its primary historical model. To understand its influence on popular discourse let us briefly consider how it operates in the theories of three seminal thinkers—Descartes himself, Francis Bacon, and Adam Smith.

RENE DESCARTES

Descartes' famous "cogito ergo sum" ("I think, therefore I exist") followed from his first methodological principle, that the beginning of all valid cognition is to prove, rather than receive and trust, one's existence. The unreliable experiential context must be doubted in every possible way on the analytical journey toward that certitude Descartes sought so urgently. More than most earlier philosophers, he reveals an almost addictive penchant for certitude or, as he put it, "clear and distinct ideas."[25]

The Cartesian premise, a radical split of method from context, operates as the conceptual core of the ideology of autonomous progress. It assumes that the scientific method generates "value-free knowledge" precisely because, and only insofar as, it is practiced in isolation from its context. The nonscientific values of the wider human culture and the personal biases of the scientist must not critique or impinge on what Mario Bunge calls "the free and lofty spirit of pure science."[26]

Within the ideology of determinism, technology shares in science's methodological power because it is nothing more than the application of science. Both the applied science position and my critique of it are complex. Put in its simplest terms, we should note that the position rests on the premise that the controlled-variable scientific method provides the unique source of objective and value-free knowledge, a premise that relegates all other forms of cognitive behavior to the status of nonknowledge; that technology holds a "middle ground" between science and nonknowledge because it applies already extant scientific knowledge to technical problems or it approaches technical problem solving with a close approximation of the scientific method.

My critique addresses two major points. On the one hand, I dispute the claim that science can ever achieve the goal of purely objective and value-free certitude even though it rightly holds to that ideal. Consequently, the claim that science represents a cognitive method radically superior to all other forms of cognition (e.g., poetry, contemplation, Aristotelian logic, storytelling, etc.) because of its freedom from personal or contextual influences cannot be sustained. In particular, I argue that *technological* knowledge is both valid and distinct from the scientific process in that it necessarily combines theoretical (and normally quantitative) models with concrete and pragmatic judgments.[27] Thus, the argument continues, just as the scientific method must not be impeded by the nonscientific bias of traditional beliefs and vested interests, so Luddite romanticism must not be allowed to hinder what Robert Heilbroner has called the "technical conquest of nature that follows one and only one grand avenue of advance.[28]

FRANCIS BACON

More commonly, progress ideology is linked with Francis Bacon's theory that sees the scientific method as guarantor of the objectivity of knowledge in the face of the deleterious effect of four "idols of the mind" that have reduced prior forms of knowledge to bankruptcy.[29] Bacon sees science as a "masculine" conqueror of nature (human nature and the larger natural order). Thus, for Bacon, science and nature do not relate to each other in a kinship of knower and known or, in the language I have been using here, of method and context. Baconian science controls and dominates nature. The following text from his "Temporis Partus Masculus" has become a locus classicus for the Baconian position.

> I come in very truth leading to you Nature with all her children to bind her to your service and make her your slave.[30]

Just as Cartesian cognitive theory denies any methodological role to the human context, so the Baconian interpretation of science reduces nature— that is, all knowable reality—to servitude in a scientific power relationship.

ADAM SMITH

The industrial capitalism of Adam Smith and the British Industrial Revolution embodies the same commitment, namely, that we must not interrupt or disturb the inner workings of the method—in Smith's case, the free market's invisible hand—by contextual interventions from society. Smith stands at the head of a long line of capitalist apologists who, in their starkest forms of orthodoxy, even argue the debilitating influence of aid to the poor. Poverty, according to their argument, results from personal character deficiency. In the calculus of the free-market mechanistic vision of the social order, the renewal of human creativity can only be guaranteed in society at large when the penalty for defective behavior is so severe that it serves as a driving motive for individual and competitive advancement. Thus, in his "Gospel of Wealth," Andrew Carnegie rejects indiscriminate charity for the poor because it encourages sloth.

> It is not the irreclaimably destitute, shiftless, and worthless which it is truly beneficial or benevolent for the individual [rich man] to attempt to reach and improve.[31]

More striking still, Herbert Spencer provides another example of the pure position:

> The well-being of existing humanity, and the unfolding of it into this ultimate perfection, are both secured by that same *beneficent, though severe discipline,* to which the animate creation at large is subject: a discipline which is *pitiless in the working out of good*: a felicity-pursuing law which never swerves for the avoidance of partial and temporary suffering. *The poverty of the incapable, the distresses that come upon the imprudent, the starvation of the idle, and those shoulderings aside of the weak by the strong, which leave so many 'in shallows and in miseries,' are the decrees of a large, far-seeing benevolence.*[32] (my emphasis)

Spencer's "large and far-seeing benevolence" operates as an inevitable methodological force (the "law" of the survival of the fittest) that, like the scientific and technological methods, must not be hindered by such contextual interventions as humanitarian concern for the destitute. His severe and pitiless law may represent the rhetorical high water mark of the Cartesian-Baconian split of method from context. Stated in brief, the principle argues that mankind must relate to every possible context for scientific, technological, and business praxis as conqueror and controller, rather than with the affection and humility of kinship. This separation of man as knower, maker, and entrepreneur from his context lies at the heart of progress ideology, a belief that scientific, technological, and business methods must be allowed to run by themselves as in capitalism's maxim, "laissez-faire."

My interpretation of the ideology of progress is perhaps unorthodox in emphasis. Nannerl O. Keohane's excellent survey of the origins and literature of the concept, for example, suggests that Descartes and Bacon provided the catalyst for gathering one set of earlier Greek and Judeo-Christian ideas into a progressive ideology. I agree and, indeed, I agree with Keohane's focus on the critical link between knowledge and control over nature. We diverge when Keohane identifies their commitment to the radically new method with "vigorous optimism."[33] It is clear that the 17th and 18th centuries were marked by an extraordinary expansion of activity in global colonization, organized scientific research, mechanized production technologies, and the birth of British and American representative government. Nevertheless, at the heart of this outburst of creativity, the Cartesian, Baconian, and capitalist laissez-faire principles sowed the seeds of doubt about the capacity of humans to interact with and benefit from their context. Indeed, Descartes and Bacon understood their first methodological task as that of overcoming the bankruptcy of the human intellectual endeavor. We have already noted Bacon's critique of the four idols of the mind. In Steven Goldman's memorable expression, "Bacon's most original contribution to Western thought" is his argument that "proper method . . . is the sole route to truth and proper method is impersonal, must be impersonal, because the single greatest obstacle to the achievement of true knowledge of nature is the activity of the human mind." Richard H. Popkins's discussion of Descartes makes a similar point.

> One finds that Descartes himself expressed great concern with the scepticism of the time; that he indicated a good deal of acquaintance with the Pyrrhonian writings, ancient and modern, that he apparently developed his philosophy as a result of being confronted with the full significance of the *crise pyrrhonienne* in 1628–29, and that Descartes proclaimed that his system was the only intellectual fortress capable of withstanding the assaults of the sceptics.[34]

Like the early 20th-century Progressive movement in the United States, this European philosophical tradition seeks security from methodological expertise rather than in the rough-and-tumble interaction of the political order. Although hardly noticed at the time, the laissez-faire premise fosters a hothouse environment aimed at protecting the workings of method from the turbulence of the larger context in which it operates. So important is this concept for understanding our 20th-century struggle with passivity, that it merits further exploration.

Three Modes of Context

The split of method from context operates on at least three levels; "context" can be understood as cultural, as natural, and as personal-affective. Let us consider each in turn.

PROGRESS AND CULTURE

Progress talk implies that scientists, technologists, and business leaders must be freed from the often well-meaning but ultimately foolish meddling of extramethodological critique. An example from American labor history may shed further light on the matter. For historian Herbert Gutman, American factory workers, beginning as early as 1840, found their "preindustrial" experience of work (as one part of life's larger cultural fabric—religion, family life, leisure, etc.) confronted by the rigors of the new factory system. Because they severed the connection between work and the rest of life, factory work rules became the battle ground for more than a century of labor-management conflict. Workers continually tried to adapt industrial work rhythms to life rhythms. Religious feasts, weddings, funerals, and other community events interrupted the regularity of factory life fostering a continual lament about absenteeism by factory masters. More important still, workers insisted on interrupting the regularities of daily work with frequent breaks for food, drink, and play.[35]

This tension echoes and perhaps even stems from a phenomenon at once physiological and psychological, the contrast between reciprocal and rotary motion. The human organism is more at home with reciprocal motion with its alternating bursts of energy and moments of rest; organic energy releases in pulses rather than in a steady stream. Rotary motion, on the other hand, never stops. It is far the more efficient mode of machine power delivery because it avoids the wear and tear of stop-start reversals found in reciprocal motion.[36] Thus, for factory machines rotary motion is the ideal. Insofar as workers are seen as functional components of the machinery, they are expected to operate in rotary-motion style, yielding steady, uninterrupted output. But insofar as workers see themselves as human organisms, they seek a reciprocal style, interrupting bursts of activity with breaks. In other words, the factory system tends to isolate work, we might call it "method," from the context of the rest of life while workers tend to insert work into a single cultural fabric.

We can apply this labor history example as a metaphor for the larger question at hand. Like factory machines, progress ideology sees "Technology" as a single moving force whose optimal mode is unencumbered expansion. From this perspective, interventions from outside the technological dynamic are unfortunate interruptions, hindrances to the inevitable forward movement.

This concept retains a vigorous presence in contemporary discourse. Consider, for example, the following text found in a recent advertisement for the United Technologies Corporation.[37] It reflects the pure ideology of autonomous progress so elegantly that it is worth a close look.

> Ethically, technology is neutral. There is nothing inherently either good or bad about it. It is simply a tool, a servant, to be refined, directed and deployed by people for whatever purposes they want fulfilled.

Note that the advertisement defines "Technology" as a single force. It is, however, an easily directed, and value-neutral tool for general human use. How, then, are we to understand technological critiques?

> So fast do times change, because of technology, that some people, disoriented by the pace, express yearning for simpler times. They'd like to *turn back the technological clock*. It is fantasy. Life was no simpler for early people than it is for us. Actually, it was far crueler. (My emphasis)

Because "Technology" equals progress, every criticism of any technology is reduced to a "longing for the primitive." Technology's unfolding dynamic comes to be identified with the passage of time itself and the reader is warned of the "utter folly" of attempting to "turn back the technological clock."

> Turning backward would not expunge any of today's problems. With technological development curtailed, the problems would fester even as the means for solving them were blunted. *To curb technology would be to squelch innovation, stifle imagination, and cap the human spirit*. (My emphasis)

"Technology," sole guarantor of human creativity, alone can solve technology's problems. The message? Don't get in its way!

> The challenge for our times is to foster *its* continued progress, to use it wisely, to manage it for *our* own greater benefit and the enrichment of life for those who follow. The full promise of technology lies in its development and use to make things *better for all*. (My emphasis)

The text ends with a pious denouement. We, whoever "we" are, must manage and use technology wisely, but not by hindering its advancement. And somehow progress will "make things better for all."

Note how the ideology of autonomous progress blurs the question of power. The identification of scientific, technological, and business methods with progress renders irrelevant the question: "Who makes the policy decisions that shape new designs?" *All changes* in science, technology, and business benefit humankind as a whole, so the question "Who wins and who loses?" has no meaning.[38]

To be precise we should consider the split between method and context less as an absolute separation than as a one-way causal flow. Method (science, technology, or business) is understood to influence its social context for the better. It is only when elements from that context seek to influence the direction of the method that the disjunction is invoked. We find a similar pattern in the relationship between science, technology, and nature.

NATURE AND PROGRESS

Carolyn Merchant's *The Death of Nature* traces the gradual shift in Western thought from a concept of Nature as bountiful goddess who provides for human needs while setting technological boundaries, to nature seen partly as a female to be conquered and partly as a passive array of resources to be exploited.[39] In recent centuries and to Western eyes, nature is rarely experienced as an interactive peer, and certainly not as a goddess. We find a partial exception, perhaps, in some 19th-century American engineers who saw "the wilderness" as a godlike force—admired, respected, dangerous, beautiful, and savage—even as they tried to conquer her.[40]

In fact, a theme of reverent affection for material reality runs through the technical traditions of the West operating in direct competition with the "method-is-all" mentality. Nevertheless, we see repeated signs of a twentieth-century technological style in the United States that preempts nature as a negotiating partner, where technical method dominates natural context. Thus, Los Angeles, unlike major nineteenth-century cities, is situated in a nearly total desert. The city's location rests on the belief that our technical systems allow us to transcend the water-constraints of the natural context. Air conditioning, which enables us to ignore the weather as long as we have enough energy to run our systems, provides a similar example. Most striking of all, of course, are our waste disposal and conservation styles, which now present us with a century of unpaid bills in such areas as acid rain and toxic pollution.

PROGRESS AND HUMAN AFFECTIVITY

The domain of progressive method is not limited to the realms of nature and culture, but extends into the affectivity of those who practice the method themselves. Scientific and technical practitioners are supposed to work their methods independently of their own affectivity if science and technology are to operate with value-free objectivity. Personal bias must not influence the work. Once again, however, the method-context split operates in a one-way direction. Human affectivity can be analyzed by social science, programmed by advertising, and reshaped by the cultural changes resulting from adaptation to technological advance.

Recently, at a colloquium between humanist scholars and scientists from the Lawrence Livermore Laboratories, I encountered an example of the sensitivity of technical practitioners to questions of personal motivation. During the day-long discussion of the morality of nuclear deterrence, one participant asked how the biases of the technicians influenced their choices of research priorities. These were scientists who claimed to judge each weapon system according to its merits as part of the deterrence calculus. Might they push lines of research simply because of their innate enthusiasm for a given project or, worse, because they wanted to keep research going to safeguard their positions? To these ears, at least, the most outspoken Livermore scientists found the question itself distinctly uncomfortable. To suggest, as the question did, that their practice of science and technology might be influenced by nontechnical motives seemed to be understood as something close to an attack on their professional integrity.[41]

This exploration of the method-context disjunction, so central to the autonomous progress model, complements our earlier observations about

America's shift from an active to a passive human participant in the scientific, technological, and business endeavors. Seen from these dual perspectives the six-word motto of the Chicago Exposition is transformed from a puzzling Depression-era slogan into an extraordinarily deft summation of the long tradition of autonomous-progress mythology. "SCIENCE FINDS" the truths of nature. "INDUSTRY APPLIES" the truths that science has found; and the term industry combines both technological and business practice. The human beings who practice the methods of all three are, as we have seen, not understood to direct their disciplines. Their proper role, like that of the larger human culture and of nature, is passive. "MAN CONFORMS" indeed!

Critiques of Progress Talk

Thus far, I have frequently observed that the "progress talk" with which we are concerned does not include all discussion, indeed very little serious academic discussion, of the value or cumulative effect of Western scientific, technological, or business practice. I have been concerned, rather, with that element of such discourse that implies that these three methodological traditions operate as inevitable forces to which human individuals, human societies, and nature must conform. Before turning to some concluding suggestions for a more appropriate language, however, it will be helpful to gather together the critiques implicit throughout the previous discussion in order to explain, explicitly, why I see such progress talk as a destructive impoverishment of our language.

In what follows I will criticize the rhetoric of autonomous progress, and the philosophical disjunction of method and context on which it depends, on three counts. I will suggest, first, that progress talk distorts historical reality; second, that it fails on simple logical grounds; and third, that it operates as a destructive political force in society, both by legitimating several centuries of Western colonialism and by its now-familiar tendency toward political passivity.

Historical Critique

Progress talk fails, first, on historical grounds. Historians of science and technology do not deny the existence of an ideal of scientific and engineering objectivity, the attempt to put one's work in the service of one's discipline and in the process to rise above mere self-interest. Even so, the claim that these methods operate in some historically abstract universe, generating value-free knowledge or value-free technical designs simply cannot be verified by the historical record.

Thus, for example, metallurgist and historian of technology Cyril Stanley Smith argues against the rhetoric of purely rational objectivity in the practice of science:

> It is high time that scientists admit that their experience in the laboratory is an aesthetic one, at times acutely so: the arid form of presenting their results has disguised this, and their respectable logical front often makes it invisible even to a student.[42]

More radically, however, it has become a commonplace among historians of science and technology that every scientific or technical artifact embodies the values and world view of its designers and maintainers.[43] Indeed, the ideology of autonomous progress is radically antihistorical at its core. If technical designs and scientific theories advance according to an inevitable interior logic, then the historian's labor of identifying the people and institutions who were decisive at any stage of either activity, of situating those actors in the specifics of their historical context, is essentially meaningless.[44] History becomes uncritical hagiography wherein the heroes of progress are celebrated for their contributions to the inevitable advance of humanity rather than studied for their influence on the shape of these essentially human endeavors.

Logical Critique

The rhetoric fails on logical grounds as well. Thus the sentence, "Technology advances," flounders between the generality of the subject, "Technology," and the specificity of the predicate, "advances." By contrast, the statement, "Internal combustion engine design advanced, between 1900 and 1930, in terms of compression ratio," is specific. I have identified the technology as well as the criteria determining which direction is forward. When it stands alone, however, "Technology" is a general term covering all those designs by which humans solve problems. It includes artifacts as diverse as hoes and shovels, steam engines, nuclear weapons, electric light and power systems, and electric toothbrushes. To speak of "Technology" in the singular, with a capital "T," and to say that it advances can only mean that whatever is done is forward. Liking or disliking progress is hardly the issue. Once the inevitability of the advance has become embedded in my language, my judgments about its values have been rendered pointless.

Political Critique: Legitimation of Colonialism

"Progress" began in the West, a circumstance laden with destructive consequence for non-Western societies. In the name of progress, "the white man's burden" as it has often been called, citizens of the West have justified and even beatified at least two centuries of colonial exploitation, continuing into the present. Westerners justified their deeds as serving the advance of civilization; we were helping others catch up. This dynamic, most visible in cross-cultural colonialism, also operates inside industrialized countries with minorities such as blacks and Native Americans in the United States and "Luddites" in 19th-century Britain. Arthur Goldschmidt exemplifies the attitude in his description of how local culture inhibits the advancement of society through Western technology transfers.

> Technical assistance personnel find the transfer of existing technology easier in the *advanced* sectors of the dual economies of the *under-developed* world, since there is generally *no cultural barrier to be breached,* no question of resistance and receptivity. . . . The witch doctor's objection to penicillin, the landowner's rejection of agricultural machinery, the merchant importer's opposition to indigenous industry. . .have greater relative significance [than in developed countries].[45] (My emphasis)

Note the classic progress language: "advanced" and "underdeveloped." Goldschmidt's cultural barrier, "the social and economic structure," is seen as the enemy of Western progress.[46]

We see the same mentality in the following poignant statement by James McLaughlin, Indian Agent for the Lakota Sioux, in 1889:

> To put the raw and bleeding material which made the hostile strength of the plains Indian through the mills of the white man, transmuting it into a manufactured product that might be absorbed by the nation without interfering with the national digestion.[47]

Political Critique: Passivity in the Western Body Politic

Finally, the ideology of progress can be critiqued in terms of the passivity it fosters among its own mainstream constituency. Because progress advances inevitably, it is understood as a force that influences, but is not influenced by, the society in which it originates. Although the culture cannot critique scientific, technological, and business methods, these same methods will gradually analyze (science) and transform (technology and business) the culture itself. The culture must "catch up" with progress.[48]

The rhetoric of citizen passivity seen in the 1933 Chicago Exposition has proved remarkably durable over the decades of this century. We find it in contemporary advertisements, in commonly accepted descriptive categories such as the HEGIS code, on the text boards of our leading museums.[49] Illogical and ahistorical though it is, the rhetoric of autonomous progress exerts continuous pressure on our imaginations as we, citizens of the body politic, wrestle with the inherently political decisions by which we allocate our scarce resources in pursuit of our various and value-laden scientific, technological, and business projects. In the brief space available here I will suggest an alternative language, one that takes into account the cumulative impact of successful technological designs while avoiding the radical passivity and implicit cultural arrogance of the model we have considered thus far.[50]

AN ALTERNATIVE LANGUAGE

The alternative language is based on the historical reweaving of technological artifacts and their contexts. Historians of technology labor to situate each artifact within the limited, historically specific, value domains from which they emerged and in which they operate. They speak of "technologies," and not "Technology," of cultural options rather than inevitable progress. This approach attempts what history traditionally holds dear, the liberation of human beings by demythologizing false absolutes and by paying attention to the human context of change.[51]

Successful technologies happen more by choice than by fate. Human beings with their tangled motives, not abstract inevitability, decide which designs are attended to and which ignored and why the technologies found worthy of inventive and fiscal attention take the final shape they do. This maxim, the central tenet of contextual history of technology, provides a basis for understanding technological style. Because a technical design reflects the motives of its design constituency, historians of technology look to the values, biases, motives, and world view of the designers when asking why a given technology turned out as it did. Every technology, then, embodies a distinct set of values. To the extent that a technology becomes successful within its society, its inherent values will be reinforced. In this sense, every technology carries its own "style," fostering some values while inhibiting others. In the technological view of history tradeoffs abound. There is no technological "free lunch."[52]

To take the matter one step further, we might note that the men and women with access to the venture capital that successful technologies always require in their early stages tend to be the same people who hold cultural hegemony in their society. "Holding cultural hegemony" means belonging to that group of people who shape the dominant values and symbols of their society. Although they never form a single historically tidy group, as a technological conspiracy theory might suggest, they do tend to view the world from the same perspective. Consequently, we can look for a set of successful technologies that, in any relatively stable era of history, embodies the "technological style" of its society.[53] It would be oversimple to say that technological style operates as the sole cause of prevailing cultural values such as those noted in our earlier discussion of the passion for systemic controls during the first three decades of this century. The values embedded in successful technologies originate in the world view of those who design and maintain them. Still, it would be equally oversimple to say that technologies exert no influence on the values of their host society. If technological designs are not value-neutral, and this is the central premise of the approach taken here, then their very success, the many ways that their host society comes to depend on them and adapt to their constraints, operates as an amplification of their inherent values.[54]

When the discussion of technological change has been rescued from the abstract ahistoricity of progress talk, it becomes clear that technological decisions always involve power relationships. Every technical choice—to invest inventive, developmental, or entrepreneurial attention in any given design, and to bring venture capital to bear on the endeavor—allocates scarce resources within the local, national, or global body politic. Which citizens have access to the design process? Whose values are embodied and whose ignored in the systems that become economically and politically successful? Questions such as these lie at the heart of any historically valid technological discourse.

It follows, then, that how we think and talk about technology influences our political as well as our intellectual stance. Insofar as we retain the linguistic habits of progress talk, we define ourselves as passive drifters on the technological tide. By doing so we choose to split our analyses of technology off from our responsibilities as actors in the socioeconomic-political drama. This, it seems to me, is the primary justification for the detailed rhetorical analysis just concluded.

One final note. Conformity, the whipping boy of the previous analysis, is not always a bad thing. Systemic conformity serves as a societal virtue—we often call it "civility" or "civic responsibility"—whenever the members share a consensus about the goals served by the system's design. Without it, in fact, communal life in a complex, system-structured society becomes unmanageable. On the other hand, should we live in a time when our consensus about how to define the common good requires renegotiation, we must be particularly careful about our technology talk. At times like that, choosing passive conformity merely hands the task of shaping technological consensus over to others. Do we live in such an era? Some interpreters of our present circumstances, myself included, think that we do.[55] But even to raise such a question, asking whether we live in a time when conformity to existing technologies is more or less constructive civic behavior, demands a language and an epistemology that define the technological endeavor in terms of its sociopolitical, and not merely its technical dimensions. Valid technological knowledge, in short, calls for an integration of quantifiable and precise theory with that form of cognition often called "intimacy," the capacity to pay attention to and find meaning within the messy inconsistencies of the larger contextual reality.

A historically critical interpretation of technology, then, demands reading the signs of the times. Responsible technology talk fosters a language of engagement, where "Technology" is understood to be a variety of particular technologies, each carrying its own embedded values, each relating to its own unique cultural circumstance. It is a language that reweaves the human fabric, reintegrating method and context, and inviting us all, technical practitioners and ordinary citizens alike, to engage in the turbulent and marvelous human endeavor of our times.

NOTES

1. U.S. Department of Commerce, Bureau of the Census, *Historical Statistics of the United States: Colonial Times to 1970,* (Washington D.C.: 1975), vol. 1, series D 1-10. I have chosen a conservative figure. Unemployment in some industrial cities reached a level of 80%.

2. Chicago Century of Progress International Exposition, *Official Book of the Fair,* (Chicago: A Century of Progress, Inc., 1932), p. 11. The earlier quote on lighting occurs on p. 21. Later, on p. 25, we find an even more stunning encomium to the exposition's lighting wizardry: "Should you gasp with amazement as, with the coming of night, millions of lights flash skyward a symphony of illumination, reflect again that it is *progress* speaking with exultant voice of up-to-the-second advancement."
I am indebted to Lowell Tozer's "A Century of Progress, 1833–1933: Technology's Triumph Over Man," *American Quarterly* 4, No.1 (Spring 1952): 78–81, for first calling my attention to the Exposition and to Cynthia Read-Miller, curator of photographs and prints in the archives at the Henry Ford Museum and Greenfield Village, for copies of the Official Book and photographs of the iconography referred to below. For texts and photos from the Guidebook see Appendix 1.

3. Lenox R. Lohr, *Fair Management: The Story of A Century of Progress Exposition,* (Chicago: The Cuneo Press, Inc., 1952), p. 96.

4. "Official Book of the Fair," p. 11.

5. Recent examples of pure progress talk will be cited below for two reasons. On the one hand they provide texts for analysis of the details of the position I have sketched out above. On the other, they constitute evidence that the rhetorical pure position remains alive and well in popular discourse today.

6. Leo Marx, "Are Science and Society Going in the Same Direction?" Remarks prepared for the 50th Anniversary Conference of the Chicago Museum of Science and Industry, "Where are We Going?"—Critical Issues in Science and Technology, 4–5 April, 1983; idem, "On Heidegger's Conception of 'Technology' and its Historical Validity," *The Massachusetts Review* 25, no. 4 (Winter 1984): 638–52; and idem, "Does Improved Technology Mean Progress?" *Technology Review* (January 1987): 32–41, 71. Merritt Roe Smith, "Technology, Industrialization, and the Idea of Progress in America," (in *Responsible Science: The Impact of Technology on Society,* ed. Keven Byrne, San Francisco: Harper and Row, 1986, pp. 1–30).

7. Marx, "Heidegger," pp. 644–45.

8. "When they spoke of progress, as they often did, they consequently gave equal weight to human betterment (intellectual, moral, spiritual) as well as material prosperity. Without betterment, prosperity was meaningless. The pursuit of science and the development of technology doubtlessly occupied an important place in this scheme of things. But as means to larger social ends, they assumed a lesser order of magnitude in the Jeffersonian scale of values." Merritt Roe Smith, "Progress," p. 4.

9. Ibid., p. 4.

10. Thomas Ewbanks, "The World a Workshop," in *Changing Attitudes Toward American Technology,* ed. Thomas P. Hughes (New York: Harper and Row, 1975), p. 115.

11. Marx ("Heidegger," p. 649) defines "technocratic" as "a commitment to the sufficiency of technical solutions or, in other words, the belief that if the means are perfected the ends will take care of themselves." Marx and Smith both call attention to the occasional critic of unabashed technical optimism (Hawthorne, Thoreau, the later Emerson, etc.) but stress their distinct minority position in popular rhetoric.

12. See, for a few further examples, Edward W. Byrne, "The Progress of Invention During the Past Fifty Years," *Scientific American* 75 (25, July 1896): 82–83; and R. H. Thurston "The Borderland of Science," *North American Review* 150 (1890: 67–79, in Hughes, *Changing Attitudes.*

Marx's analysis of Heidegger's concept of "enframing" stresses the same aggressive quality noted in Smith's essay. Thus: "Enframing, the revealing the rules through modern technology, is far more *aggressive, intrusive, extractive, not to say rapacious* than earlier modes of revealing" (Heidegger," p. 641; my emphasis).

In what follows I will adopt the admittedly sexist use of "man" for "human" both because the usage is commonplace in the rhetoric under discussion and because it reflects the stereotypical division of gender into the separate spheres that underlies this whole mode of thought.

13. James W. Carey and John J. Quirk, "The Mythos of the Electronic Revolution," *The American Scholar,* 39, nos. 1, 2 (Spring and Summer 1970): 228, and, for the McLuhan text p. 402.

14. For an overview of scholarship on the topic, see Paul Boyer's " 'The Ragged Edge of Anarchy': The Emotional Context of Urban Social Control in the Gilded Age," in his *Urban Masses and Moral Order in America, 1820–1920* (Cambridge: Harvard University Press, 1978).

15. See John Higham, *Send These to Me: Jews and Other Immigrants in Urban America* (New York: Atheneum, 1975), for a broad overview of legal and social resistance to immigration from 1870 through the draconian 1924 immigration act that marked the definitive end of the earlier open-door policy. On the origin and later history of the popular image of "The Melting Pot," see Philip Gleason, "The Melting Pot: Symbol of Fusion or Confusion?" *American Quarterly* 16, no. 1 (Spring 1964): 20–46.

16. J. L. Larson, "A Systems Approach to the History of Technology: An American Railroad Example," a paper read at the annual meeting of the Society for the History of Technology, 1982, p. 17.

17. The most helpful single source on changing labor-management relations in 19th-century America remains Herbert Gutman's *Work, Culture and Society in Industrializing America* (New York: Vintage, 1966), esp. the title chapter.

For a more detailed study of the growth of standardization as a technological ideal, together with a correlative decline in the more political ideal of negotiation as a means of solving social problems, see my "The Politics of Successful Technologies," in Stephen Cutcliffe and Robert Post, *In Context: History and The History of Technology—Essays in Honor of Melvin Kranzberg,* Bethlehem, Pa.: (Lehigh University Press, 1988).

18. The text, with what immediately follows, is from A. Michael McMahon, "An American Courtship: Psychologists and Advertising Theory in the Progressive Era," *American Studies* 13 (1972): 6.

19. Robert Bellah, et al., *Habits of the Heart: Individualism and Commitment in American Life* (Berkeley: University of California Press, 1985), p. 261.

20. See Stephen Meyer, *The Five Dollar Day: Labor Management and Social Control in the Ford Motor Company, 1908–1921* (Albany: State University of New York Press, 1981), passim.

21. On Scientific Management and later industrial psychology movements, see Edwin T. Layton Jr., *The Revolt of the Engineers: Social Responsibility and the American Engineering Profession* (Cleveland: Case Western Reserve Press, 1971), and Samuel Haber, *Efficiency and Uplift: Scientific Management in the Progressive Era, 1890–1920* (Chicago: University of Chicago Press, 1964). David Brody's *Workers in Industrial America: Essays on the 20th Century Struggle* (New York: Oxford University Press, 1980), provides a thoughtful analysis of welfare capitalism and the pre- and postwar labor-management context. For an interpretation focusing more on radical worker movements in the period, see James R. Green, *The World of the Worker: Labor in Twentieth Century America* (New York: Hill and Wang, 1980). On the introduction of tear gas into civilian police forces, see Daniel P. Jones, "From Military to Civilian Technology: The Introduction of Tear Gas for Civil Riot Control," *Technology and Culture* 19, no. 2 (April 1978): 151–68.

22. Roland Marchand, *Advertising the American Dream: Making Way for Modernity, 1920–1940,* (Berkeley: University of California Press 1985), pp. 68–69; see also p. 10 but also passim. See also McMahon, "American Courtship," and Daniel Pope, *The Making of Modern Advertising* (New York: Basic Books, 1983). For a provocative analysis of the new advertising style as part of a larger social movement from the 19th-Century Protestant ethic of productivity toward the 20th-Century therapeutic ethos, see T. J. Jackson Lears, "From Salvation to Self-Realization: Advertising and the Therapeutic Roots of the Consumer Culture, 1880–1930," in *The Culture of Consumption: Critical Essays in American History, 1880–1980,* ed. Richard Wightman Fox and T. J. Jackson Lears (New York: Pantheon Books, 1983), pp. 1–38.

23. We should not, however, overlook the strength of the Carey and Quirk hypothesis. Electronic technologies, as Henry Adams intuitively observed in his famous "Virgin and Dynamo" essay, present a radically new, quasi-mystical technological ideal qualitatively different from the more immediately obvious and sensual technologies of the preelectronic machine age. For an elegant argument of the symbolic importance of the shift from a gear-and-girder to an electronic technical ideal, see the preface of Cecelia Tichi's *Shifting Gears: Technology, Literature, Culture in Modernist America* (Chapel Hill: University of North Carolina Press, 1987), pp. xi–xvi.

24. This year's bicentennial celebration of the Constitution serves as a vivid reminder of the founding fathers' extraordinary confidence in their ability to craft a new social order. They would, I suspect, be dismayed were they to witness the emergence of the doctrine of conformity in the 20th-century version of belief in progress.

25. Richard H. Popkin provides the broad philosophical and theological context as well as the detailed personal circumstances that render Descartes's extreme need for certitude intelligible; see *The History of Scepticism from Erasmus to Descartes* (Assen, The Netherlands: Koninklijke Van Gorcum & Comp. N.V. Assen, 1960), esp. chaps. 9, 10)

For an insightful analysis of the Cartesian assumption as it has influenced Western science, see Evelyn Fox Keller and Christine Grontkowski, "The Mind's Eye," in *Discovering Reality* ed. Sandra Harding and Merril B. Hintikka (Dordrecht, The Netherlands: D. Reidel, 1983), pp. 212, 214. Keller and Grontkowski situate the Cartesian disjunction in the much longer Western tradition of Greek philosophy. On the other hand, for a portrayal of Greek ambivalence about scientific and technological progress in the dramatic poetry of Aeschylus, Sophocles, and Euripides, see Arthur D. Kahn, "Every Art Possessed by Man Comes from Prometheus: The Greek Tragedians and Science and Technology," *Technology and Culture* 11, no. 2 (April 1970): 133–62.

For a similar interpretation of Descartes, see Hans-Georg Gadamer, *Truth and Method* (New York: Seabury, 1975), pp. 245–53. Robert Nisbet makes the same point in his somewhat cavalier treatment of Descartes; see *History of the Idea of Progress* (New York: Basic Books, 1970), pp. 115–17. For one of Michael Polanyi's frequent passing references to the disjunction, see his *Personal Knowledge: Towards a Post-Critical Philosophy* (Chicago: University of Chicago Press, 1958), p. 269. Finally, for a thorough analysis of the relationship between the "cogito" and Cartesian theory of scientific knowledge, see James Collins, *Descartes' Philosophy of Nature* (Oxford: American Philosophical Quarterly Monograph Series, 1971).

26. For this text and a detailed articulation of the autonomous science position, see Mario Bunge, "Technology as Applied Science," *Technology and Culture* 7, no. 3 (Summer 1966): 329–47.

27. Lest this overly simple sketch of my position be misunderstood, let me hasten to note that my critique of the controlled-variable method pertains only to what is sometimes called "Scientism," namely: (a) claims for its ability to overcome *all* bias and achieve a form of *absolute* certitude and (b) the consequent claim that other cognitive traditions are nonknowledge. The value of the ideal of precision and objectivity for limiting the influence of bias in scientific traditions is, I take it, obvious. It is, however, one thing to seek precision and objectivity as an asymptotic ideal and quite another to assert its absolute achievement in the face of historical evidence to the contrary. For references to some of that evidence see n. 43 below. I see science, in short, as a very welcome member of the family of human cognitive traditions but not as the one and only valid way toward truth.

Finally, for a much more thorough presentation of all this see my *Technology's Storytellers: Reweaving the Human Fabric* (Cambridge: The MIT Press, 1985), chap. 3 and esp. pp. 96–99.

28. The original Luddites were 19th-century British textile workers who selectively destroyed those new machines that they judged damaging to themselves. In progress talk the term "Luddite" is often used to stigmatize those whose fear of change leads them to reject technological advances of any sort. See David F. Noble, "In Defense of Luddism," *Democracy* 3, nos. 2, 3, 4 (Spring, Summer, Fall, 1983) but esp. no. 2 (Spring): 8–24, and Adrian J. Randall, "The Philosophy of Luddism: The Case of the West of England Woolen Workers, ca. 1790–1809," *Technology and Culture* 27, no. 1 (January 1986): 1–17.

For the text just cited, see Robert L. Heilbroner, "Do Machines Make History?" in *Technology and Culture* 8, no. 3 (July 1967): 337. For my detailed critique of Heilbroner's position, see *Storytellers*, pp. 140–43.

One further example of the commonplace description of technology as applied science can be found in the HEGIS Codes for academic disciplines. The subheading "Engineering and Technology" is defined as follows: "Includes those subject field designations associated with the practical application of basic scientific knowledge to the design, production, and operation of systems intended to facilitate man's control and use of his natural environment." Note that, in addition to the expression "application of basic science," the HEGIS classification adopts other code words from the ideology of autonomous progress, in particular "*control . . . of his* natural environment."

29. The four idols are: (a) "idols of the tribe" (the tendency to stress only that evidence that affirms one's theory); (b) "idols of the cave" (the personal bias of each person); (c) "idols of the market-place" (the biases embedded in ordinary and

unquestioned language); and (d) "idols of the theater" (the biases inherited from our educational training). For an illuminating discussion of the overall position as well as a contrast with the Renaissance nature philosophy against which Bacon fought, see Steven L. Goldman's "From Love to Gravity: Renaissance Nature Philosophy versus Modern Science," unpublished manuscript, Lehigh University, Bethelehem, Pa. Or see James Collins *A History of Modern European Philosophy* (Milwaukee: Bruce, 1954), pp. 54–57.

30. Francis Bacon, "Temporis Partus Masculus: An Untranslated Writing of Francis Bacon," in B. Farrington, *Centaurus* 1 (1951): 197. Cited in Evelyn Fox Keller, "Baconian Science: A Hermaphroditic Birth," *The Philosophical Forum* 11, no. 3 (Spring 1980): 30.

31. Edward C. Kirkland, ed., *The Gospel of Wealth and Other Timely Essays by Andrew Carnegie,* (Cambridge: Belknap Press of Harvard University, 1962), p. 31.

32. "The Sins of Legislators," in *The Man Versus the State* 1892; reprint, (Indianapolis: Liberty Classics, 1981), pp. 107–108. See also in the same work, "The Coming Slavery," esp. pp. 33–34.

For a discussion of the link between the ideology of autonomous progress and Adam Smith's invisible hand, see Langdon Winner, *Autonomous Technology: Technics-out-of-Control as a Theme in Political Thought* (Cambridge: The MIT Press, 1977), pp. 293–94, and Richard E. Sclove, "Energy Policy and Democratic Theory," in *Uncertain Power: The Struggle for a National Energy Policy,* ed. Dorothy S. Zinberg (New York: Pergamon, 1983), pp. 52–53.

33. See Nannerl O. Keohane, "The Enlightenment Idea of Progress Revisited," in *Progress and Its Discontents,* ed. G. A. Almond, M. Chodorow, and R. H. Pearce (Berkeley: University of California Press, 1977), pp. 21–40. For other helpful overviews of the European progressive tradition, see Nisbet, "History of the Idea of Progress," and Howard Segal, *Technological Utopianism in American Culture* (Chicago: University of Chicago Press, 1985), chap. 4 This is also discussed in Goldman's "Science, Philosophy and Religion in the 17th Century," in *The Impact of Science and Technology on Society and Religion,* ed. Joseph J. Kockelmans (Pennsylvania State University Press, n.d.).

34. Steven Goldman, "From Love to Gravity," p. 5. Popkin, "History of Scepticism," p. 175.

35. Herbert Gutman, "Work, Culture, and Society in Industrializing America," in his book by the same title (New York: Random House, 1966).

36. I am indebted to the late Lynn White Jr. for this observation about the nonorganic quality of rotary motion. See his *Medieval Technology and Social Change* (New York: Oxford University Press, 1962), p. 115.

37. *Harpers* (March 1983): 3. For a copy of the complete text see Appendix 2.

38. For a more complete discussion of the question of power relationships in technological decision making, see my "The Politics of Successful Technologies."

39. Carolyn Merchant, *The Death of Nature: Women, Ecology, and the Scientific Revolution* (New York: Harper & Row, 1980).

40. See, for example, Elting E. Morrison, "The Works of John B. Jervis," in his *From Know-how to Nowhere: The Development of American Technology* (New York: Basic Books, 1974).

41. Colloquium on "The Ethics of Nuclear Deterrence," 14 February 1986. Bishop John S. Cummins, of Oakland, California, hosted the conference. Participants came primarily from the Lawrence Livermore Laboratories, the University of California at Berkeley, and the Graduate Theological Union of Berkeley.

42. Cyril Stanley Smith, "The Cover Design: Art, Technology, and Science: Notes on their Historical Interaction," *Technology and Culture* 11, no. 4 (October 1970): 498.

43. Such critiques pervade recent history of science where the claims of objective and value-free science are challenged by studies of the political and economic factors influencing the content and methodology of science. See, for example, Arnold Thackray, "The History of Science," in *A Guide to the Culture of Science, Technology, and Medicine*, ed. Paul Durbin (New York: Free Press, 1979), and Roy MacLeon, "Changing Perspectives in the Social History of Science," in *Science, Technology, and Society: A Cross-Disciplinary Perspective*, ed. Ina Spiegel-Rosing and Derek De Solla Price (Beverly Hills, Calif.: Sage, 1977). For studies of psychological and philosophical influences on the content and method of science, recent "gender and science" literature is particularly important. Evelyn Fox Keller has developed this perspective most thoroughly and insightfully. See her *A Feeling for the Organism: The Life and Work of Barbara McClintock* (San Francisco: W. H. Freeman, 1983), and *Reflections on Gender and Science* (New Haven: Yale University Press, 1985).

For an overview of similar critiques concerning the influence of bias on research and development and technological design in the history of technology see my *Storytellers,* passim but esp. chap. 1 and 5. For a gender and technology perspective, see Joan Rothschild, ed., *Machina ex Dea: Feminist Perspectives on Technology* (New York: Pergamon Press, 1983), esp. her introduction.

44. Arnold Gehlen demonstrates the point in the following passage. "Technology [can hardly be seen as] the product of a *conscious* human effort to extend material power, but rather as a *large-scale biological process* by which the human organism's innate structures are impressed onto the human environment to an ever greater extent: a biological process, in other words, which because it is just that, is *beyond the reach of human control.* Both types of process, technological progress and biological development under the pressures exerted by the industrial system, have entered a phase of incalculable endlessness, and this alone compels us to call the era in which we live . . . *a post-historical phase.* "Über kulturelle Evolution," cited in Reinhard Rurup, "Historians and Modern Technology: Reflections on the Development and Current Problems of the History of Technology," *Technology and Culture* 15, no. 2 (April 1974): 164.

45. Arthur Goldschmidt, "Technology in Emerging Countries," *Technology and Culture* 3, no. 4 (Fall 1962): 587–89.

46. For a study of the influence of this Western bias on the research preferences of mainstream historians of technology, see *Storytellers,* pp. 26–34, and 178–79.

47. Source unknown. The text appears as part of a photographic exhibit of the Lakota people by Donald Doll, S.J.

48. See, for one of the more noted examples of this position, William Fielding Ogburn's theory of "cultural lag" in his *Social Change with Respect to Culture and Original Nature* (New York: Huebsch, 1923). For a favorable assessment of the influence of Ogburn's theory, see F. R. Allen, "Technology and Social Change: Current Status and Outlook," *Technology and Culture* 1, no. 1 (Winter 1959): 48–59. For a critique, see Reinhard Rurup, "Historians and Modern Technology," p. 188.

49. Thus, for example: "The same technology can be used to launch satellites or warheads. Technology itself has no moral dimensions. It is neither good nor evil. The use to which technology is put is determined by mankind for political, economic, and social reasons" (wall plaque in Space Hall next to the Jupiter C, Scout D, Minuteman III, and Vanguard missiles in the National Aerospace Museum, Washington, D.C.). On the HEGIS code, see note 28 above.

50. For the sake of simplicity I will omit further treatment of the ideology of progress as it relates to science and business and concentrate exclusively on technological designs.

51. See my "Storytellers" (chap. 5) for evidence that historians of technology have succeeded in this contextual interpretation of technological change and for counterevidence that they themselves been influenced by the ideology of autonomous technological progress in their choice of research subjects.

52. On the tradeoffs inherent in technological style, see Thomas P. Hughes, "We Get the Technology We Deserve," *American Heritage* (October–November 1985): 65–79.

53. On cultural hegemony, see T. J. Jackson Lear's "The Concept of Cultural Hegemony: Problems and Possibilities," *The American Historical Review* 90, no. 3 (June 1985): 567–93.

54. For a more complete discussion of technological style and the entire model proposed here, see my *Storytellers,* pp. 192–201 and "The Politics of Successful Technologies."

55. See, for example, Bellah et al., *Habits of the Heart.* The authors present compelling evidence of America's individualistic incapacity for and need to renegotiate a vision of the common good. I have argued the same point in my "United States Technology and Adult Commitment," *Studies in the Spirituality of Jesuits* (St. Louis: Seminar on Jesuit Spirituality, 19, no. 1, January 1987).

Appendix 1
The Chicago 1933 Exposition
Photographs and Texts
THEME OF FAIR IS SCIENCE

As two partners might clasp hands, Chicago's growth and the growth of science and industry have been united during this most amazing century. Chicago's corporate birth as a village, and the dawn of an unprecedented era of discovery, invention, and development of things to effect the comfort, convenience, and welfare of mankind, are strikingly associated.

Chicago, therefore, asked the world to join her in celebrating a century of the growth of science, and the dependence of industry on scientific research.

An epic theme! You grasp its stupendous stature only when you stop to contemplate the wonders which this century has wrought.

Science Finds—Industry Applies—Man Conforms

Science discovers, genius invents, industry applies, and man adapts himself to, or is molded by, new things. Science, patient and painstaking, digs into the ground, reaches up to the stars, takes from the water and the air, and industry accepts its findings, then fashions and weaves, and fabricates and manipulates them to the usages of man. Man uses, and it effects his environment, changes his whole habit of thought and of living. Individuals, groups, entire races of men fall into step with the slow or swift movement of the march of science and industry.

There, in epitome, you have a story that A Century of Progress tells you, not in static, lifeless exhibits, but in living, moving demonstrations of beauty and color. Science, to many of us, has been only a symbol of something mysterious, difficult, intricate, removed from man's accustomed ways. So few of us realize that in virtually everything that we do we enjoy a gift of science. A Century of Progress undertakes to clothe science with its true garb of practical reality and to tell its story of humanly significant achievement so that even he who runs may read.

Illustration of the Administration Building with two sculpture figures symbolizing science and industry, Chicago World's Fair, 1933, from catalog, "Chicago and Its Two Fairs, 1893–1933" (Chicago: Geographical Publishing Co., 1933). *From the collections of Henry Ford Museum and Greenfield Village.*

"The Fountain-Hall of Science," photograph by Kaufman & Fabry Co., in booklet, "Official World's Fair in Pictures: A Century of Progress Exposition—1933" (Chicago: The Reuben H. Donnelley Corp., 1933). *Courtesy of The Henry Ford Museum, Dearborn, Michigan.*

Appendix 2
United Technologies

(Reprinted with permission, see note 37)

TECHNOLOGY'S PROMISE

Know-how applied to get a job done—that's technology. The job can be as simple as toasting a slice of bread or as complex as mounting an odyssey to the planets.

There's nothing simple, of course, about the tasks on society's agenda today. Contemporary needs and challenges are the complex concomitants of a complex world. And meeting them will demand technologies more advanced, and with greater capabilities, than any that have gone before.

Ethically, technology is neutral. There is nothing inherently either good or bad about it. It is simply a tool, a servant, to be refined, directed and deployed by people for whatever purposes they want fulfilled.

In its best uses, technology is a means for solving problems and achieving objectives. It helps to improve life. It lets people do more things in better ways. Throughout history, it has functioned as the cutting edge of human progress. With his technology, man has overcome uncounted problems that burdened him for centuries.

Life as we know and want it would be impossible without modern technology—in agriculture, industry, communications, health care, transportation, housing, and a thousand other spheres.

So fast do times change, because of technology, that some people, disoriented by the pace, express yearning for simpler times. They'd like to turn back the technological clock. But longing for the primitive is utter folly. It is fantasy. Life was no simpler for early people than it is for us. Actually, it was far crueler.

Turning backward would not expunge any of today's problems. With technological development curtailed, the problems would fester even as the means for solving them were blunted. To curb technology would be to squelch innovation, stifle imagination, and cap the human spirit.

Technology to serve people must continue to be nurtured and strengthened and honed and put to the tasks in society that need to be done.

The challenge of our times is to foster its continued progress, to use it wisely, to manage it for our own greater benefit and the enrichment of life for those who follow. The full promise of technology lies in its development and use to make things better for all.

—United Techologies